CARBON CAPTURE, STORAGE, AND UTILIZATION

CARBON CAPTURE, STORAGE, AND UTILIZATION

A possible climate change solution for energy industry

Malti Goel • M Sudhakar • R V Shahi

CRC Press
Taylor & Francis Group
Boca Raton London New York

CRC Press is an imprint of the
Taylor & Francis Group, an **informa** business

The Energy and Resources Institute

CRC Press
Taylor & Francis Group
6000 Broken Sound Parkway NW, Suite 300
Boca Raton, FL 33487-2742

First issued in paperback 2023

© 2018 by Malti Goel, M Sudhakar, R V Shahi and The Energy and Resource Institute

CRC Press is an imprint of the Taylor & Francis Group, an informa business

ISBN 13: 978-1-03-265389-1 (pbk)
ISBN 13: 978-0-367-17908-3 (hbk)
ISBN 13: 978-0-429-05911-7 (ebk)

DOI: 10.1201/9780429059117

Print edition not for sale in South Asia (India, Sri Lanka, Nepal, Bangladesh, Pakistan or Bhutan)

Library of Congress Cataloging in Publication Data
A catalog record has been requested

Visit the Taylor & Francis Web site at
http://www.taylorandfrancis.com

and the CRC Press Web site at
http://www.crcpress.com

Foreword

Carbon Capture and Storage (CCS) has emerged as one among the three key energy technology options to mitigate climate change. The other two are: energy efficiency improvement and increasing use of renewable energy sources.

Introduced in 1990s, the CCS technology is seen as an option to address problems of climate change, and has also added a new dimension to scientific research in addressing one of the most contentious issues of the 21st century. The other two technology options are already being pursued in a mission mode under National Action Plan on Climate Change. CCS being a late entrant to the scene, has attracted the attention of both researchers and policy makers.

The research in carbon sequestration took shape in the Department of Science and Technology under its Inter-sectoral programmes. I am happy that this book on Carbon Capture, Storage, and Utilization (CCSU) edited by Malti Goel, M. Sudhakar, and R. V. Shahi is a unique text about the options for CO_2 fixation. A large number of research frontiers have been covered in the book. The need of the hour is for the Industry and Government Departments to pool their resources to address CCSU-related issues. The policy paper of Shri R. V. Shahi, Ex-Secretary Power, in this context is of immense value.

We need to take steps to tackle climate change arising from economic growth. Energy is the engine for economic growth. The question is, could CCS become a turning point in the future in the energy sector and what we can do in this emerging field? The book discusses potential of an innovative energy technology CCSU towards a sustainable energy future. Various techniques of CO_2 recovery from power plants by physical, chemical, and biological means as well as challenges and prospects in biomimetic carbon sequestration are addressed in this book. It showcases Indian research perspectives to the world and that makes it a significant contribution. It is sincerely hoped that this book will be a valuable guide for policy makers and serve as a reference book also.

Harsh Gupta
Member, Atomic Energy Regulatory Board
President, International Union of Geodesy and Geophysics
President, Geological Society of India

Message

I have long association with Coal India Ltd., a company which produces so much of coal. We are the third largest coal producer and the Coal India is the single largest coal producing company in the world. I personally believe that by producing coal, their responsibility does not cease. Something beyond that must be pursued for reducing its carbon footprints. On the research and development (R&D) activities for carbon sequestration, Coal India's contribution till today is almost negligible. I think that one of the cash-rich companies like Coal India with around 42,000 crores cash liquidity has a role to play and should invest in R&D and technology development.

I fully agree with Dr. Kasturirangan, Former Member, Planning Commission, that we would need to find implementable solutions for carbon capture and storage (CCS) through continuous R&D activity and in this area capacity building is important. In this context, contribution of Dr. Malti Goel, Former Adviser and Scientist 'G' in the Department of Science & Technology, who conceived and spearheaded the research under National Programme on CO_2 Sequestration, is very much appreciated. I think institutions like Indian School of Mines, Banaras Hindu University and other mining institutions and universities should pay serious attention to it. There is a need to give thrust to R&D and also special courses have to be conducted. At present about 600 million tonnes of coal is consumed and it would soon become 1.0 billion tonnes and then 1.6 billion tonnes. Once you capture carbon dioxide, the variant is how to store it. That is going to be a real challenge. I would suggest that to sequester CO_2 inside the mine need to be given a push. While industries like NTPC, ONGC and Reliance are coming forward to invest in R&D, I am sure that Coal India will also involve themselves in these activities.

Recently, Hon'ble Minister of State with Independent Charge for Power, Coal and New & Renewable Energy, Government of India Shri Piyush Goyal has said that top priority is to be given to R&D projects in Energy sector. I feel extremely happy that an edited volume of the lecture notes of capacity building workshop on carbon capture and storage:

earth processes organized by Climate Change Research Institute is being published as *Carbon Capture, Storage and Utilization* by TERI Press. This book certainly would form a noteworthy resource for future researchers as well as compose an educational material on the subject.

M. P. Narayanan
Former CMD, Coal India Ltd

Acknowledgement

India is the third largest coal producer and coal consumer country, accounted for about 8% of the world coal consumption in 2012. The climate change concerns arising from coal combustion and increasing accumulation of greenhouse gases have given rise to the need for development of clean energy technologies; and CO_2 sequestration is one of them. This book, *Carbon Capture, Storage, and Utilization: a possible climate change solution for energy industry*, is about CO_2 Sequestration technology.

In this endeavour our whole hearted thanks are to Hon'ble Dr Kasturirangan, Member, Planning Commission; and to Vice Chancellor, Jawaharlal Nehru University for his blessings and encouragement; and to Prof. H. K. Gupta, Member, National Disaster Management Authority, who has been extremely gracious in giving his valuable time, advice and guidance.

We are indebted to Dr Shalaish Nayak, Secretary, Ministry of Earth Sciences for the valuable guidance, support and interactive discussions; and to Prof. A. K. Ghoshal, IIT Guwahati and Prof. T. Satyanarana, Delhi University South Campus for their active involvement and unstinted cooperation.

We feel obliged to Dr M. P. Narayanan Ex-CMD, Coal India Ltd and Shri D. K. Aggrawal, Executive Director, National Thermal Power Corporation for giving industry perspectives and recommendations about taking up pilot projects.

The Climate Change Research Institute in association with the Ministry of Earth Sciences and other stakeholders organized the workshop on Carbon Capture and Storage: Earth Processes at India International Centre (IIC) during 15–19 January 2013. This publication contains a compilation of lectures delivered in the workshop and also those which were proposed but could not be delivered, to widely disseminate the knowledge on the subject.

We acknowledge contributions made by executive committee members of the Institute and the IIC in successful organization of the capacity building workshop.

We are very much thankful to Ms. Anupama Jauhary, Head - TERI Press, and her team for bringing out this publication in the present form.

Every accomplishment requires hard work and good wishes of many people. We thank all those who are not mentioned.

Introduction

This book is an outcome of the workshop on "Carbon Capture and Storage: Earth Processes". Held from 15 to 19 January 2013, it was supported by Ministry of Earth Sciences, Government of India and organized by Climate Change Research Institute. The aim of this edited book is to analyse how current research on carbon capture, storage and utilization (CCSU) is being pursued in the world and in India, and what possible implications it may have in finding solutions to pollution problems from the energy sector.

Environmental pollution to a large extent is linked to energy, which is the engine for economic growth. Energy from fossil fuels combustion is giving rise to increased emissions of carbon dioxide (CO_2), a greenhouse gas (GHG). CO_2 and other natural GHGs from anthropogenic activities are responsible for maintaining the global average temperature at 14.43°C. Increased emission of GHGs and their accumulation in the atmosphere are affecting the earth's average temperature, and giving rise to global warming and climate change. In 2010, around 1852.45 kg of oil equivalent (kgoe) of energy was consumed by the world. Figure 1 shows the world's electricity generation in fuel. The contribution of CO_2 and GHGs to global warming is shown in Figure 2.

In response to Intergovernmental Panel on Climate Change (IPCC) 2007 assessment report that global average temperature having risen by 0.74°C in hundred years from 1906 to 2005, Conference of the Parties (COP) to the United Nations Framework Convention on Climate Change (UNFCCC) decided in 2009 that actions should be taken to limit the increase in global average temperature to 2°C. The IPCC 2013 assessment report has further made it unquestionable and concluded that human activities are continuing to affect earth's energy budget and warming from 1880 to 2012 can be estimated as 0.85°C.

Climate change offers an opportunity to discover better understanding of our planet and role of excess energy consumption in anthropogenic activities. Energy-climate change modelling studies' predictions are that with increasing consumption of energy, the CO_2 concentrations are

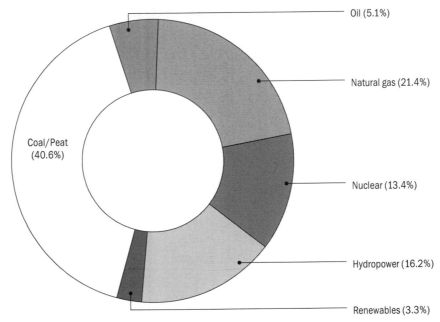

Oil (5.1%)

Natural gas (21.4%)

Coal/Peat
(40.6%)

Nuclear (13.4%)

Hydropower (16.2%)

Renewables (3.3%)

Figure 1 *World energy generation by resource in 2010*
Source: International Energy Agency (IEA), Key World Energy Statistics Paris: IEA, 2011. http://www.iea.org/ Textbase/ publications/free new_ Desc.asp?PUBS_ID=1199

expected to rise to 550 ppm (parts per million) in 2050. To achieve 2°C limit, the CO_2 concentration should not be allowed to increase beyond 450 ppm and this has been set as a goal for saving the planet from any climate-induced calamity. The CO_2 current level is 400 ppm as against the pre-industrial value of 280 ppm.

It is in this context that environment, climate change, and development issues are getting prominent among the global populations and the availability of energy resources and energy technologies are becoming the key concerns. Shackley and Duitshke[1] put forward a *CCS Rational Pyramid* with a wider base of scientific knowledge about climate change and the need for making deep cuts into CO_2 concentration at the bottom of the pyramid. Various technology options and generating systems are discussed. CCS projects occupy top position in the pyramid. The CCS technology is an emerging technology currently in demonstration phase. By its application, relative increase in levelized cost of electricity in a

---

[1] Shackley, Simon, and Elisabeth Duitshke. 2012. Introduction to special issue on carbon dioxide capture and storage – Not a silver bullet to climate change but a feasible option. *Energy & Environment* 23(2&3):209–225

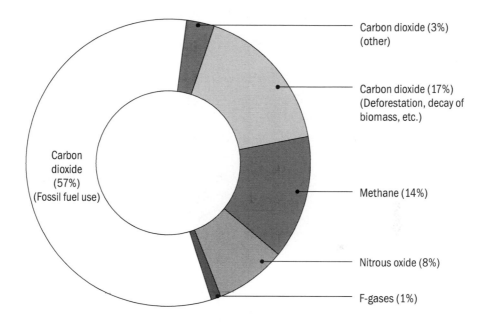

Figure 2   *Contribution of various GHGs to global warming*
Source: IPCC (2007); based on global emissions in 2004. Details about
the sources included in these estimates can be found in the Contribution
of Working Group I to the Fourth Assessment Report of IPCC

supercritical power plant with CO_2 capture can be as high as 63% in comparison to a plant without CCS.[2]

World Energy Outlook[3] report of IEA has pointed out that "nearly four-fifths of the total energy related CO_2 emissions permissible by 2035 in the 450 scenario are already 'locked-in' by our existing capital stock." This would mean that future power plants (post-2020) should be zero emission plants. Further, outlook for the current share and growth projections of various technologies in world electricity generation for CO_2 mitigation in the atmosphere is as follows:

Renewable energy: Under the projected growth scenario, the share of renewable in electricity is expected to increase from 3% in 2009 to 15% in 2035. Renewable energy remains the most favoured option. The projection also made for hydropower generation to global power continues to remain at the present level, that is, about 15%.

---

[2] Kulichenko, Natalia, and E. Ereina. 2012. *Carbon Capture and Storage in Developing Countries: A Perspective on Barriers to Deployment*. Washington DC: World Bank

[3] World Energy Outlook, International Energy Agency, http://www.iea.org/

Coal-based generation: Realization of 450 ppm scenario requires that the world's coal consumption declines from 2020 onwards. However, the projections are that coal use is expected to rise and that demands in 2035 will nearly double the total world's coal demand of 2009.

Nuclear power: In nuclear power, there is no significant decline expected even after Fukushima nuclear disaster of 2011. On the contrary, nuclear output is projected to rise more than 70% over the period from now to 2035.

Energy efficiency: Energy efficiency increases in both generation and demand sectors. The increase in efficiency of coal-based power plants by 5% in 2035 can lower CO_2 emissions by as much as 8%. Energy efficiency in demand sectors would contribute much larger (20%–25%) reductions in emissions.

Carbon capture and storage (CCS):   The CCS plays an important role as a key abatement option in post-2020, amounting to one fifth of reductions in the CO_2 concentrations by 2050.

Model analysis of IEA Energy Technology Perspectives 2010 can be cited to develop pattern of economic growth and determine technological choices for a sustainable energy future. Two objectives for scenario analysis of future technology systems are Accelerated Technology (ACT) map, which envisages bringing global energy-related CO_2 emissions in 2050 to same as 2005 level, and Blue map, which envisages the global energy-related CO_2 emissions in 2050 to become half of the emissions in 2005. From a number of scenario options under ACT and Blue map, the policy instruments, price trajectory, and penetration of different technologies including CCS have been devised. In ACT map scenarios, end-use energy efficiency provides most emission reductions. However, the growth of CCS, from ACT to Blue map scenarios, could result in 32% of the additional emissions reduction. The CCS is therefore interlinked to the way we use energy resources and goals set for saving the planet. Future projection of global deployment of CCS from high purity CO_2 sources during 2010–50 is shown in Figure 3.

Having projected the vital role of CCS in future energy systems, several international organizations and multi-lateral bodies have taken upon themselves to promote CCS technologies. A large number of pilot and demonstration projects on CCS have been launched worldwide. In the USA, projects to develop CO_2 capture technologies and to demonstrate storage and fuel recovery under FutureGen, Weyburn, Frio, Pilot Allison

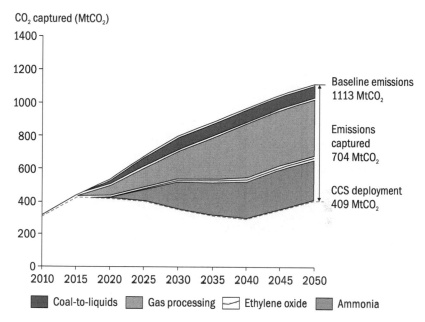

CO$_2$ captured (MtCO$_2$)

Baseline emissions
1113 MtCO$_2$

Emissions
captured
704 MtCO$_2$

CCS deployment
409 MtCO$_2$

■ Coal-to-liquids   ▨ Gas processing   ▭ Ethylene oxide   ■ Ammonia

Figure 3   *Global deployment of CCS from high purity CO$_2$ sources 2010–50*
Source: http://www.globalccsinstitute.com/node/42436

and others are in progress in gas-fired and coal-fired plants. Sliepner saline aquifer project was first to begin in 1996 in Norway for capturing CO$_2$ from Statoil natural gas field and storage under North Sea. In European Union many other collaborative projects such as CO$_2$ SINK, Snohvit, In Salah have since begun to test the reliability and safety studies of CO$_2$ storage. In the UK, plans to contract large infrastructure projects around five coal-fired power stations for CO$_2$ storage under North area are in pipeline. Australia's research focus is on large-scale CCS demonstrations project by CO$_2$ CRC (Cooperative Research Centre for Greenhouse Gas Technologies) and Otway project which began injection in April 2008 with continuous monitoring and verification studies. People's Republic of China (PRC), Saudi Arabia, Brazil including India are other countries planning research and development (R&D) on CCS. Main goals of these projects can be summarized as: (a) feasibility of large-scale operation by reducing the cost of capturing CO$_2$ and (b) testing the efficacy of CO$_2$ fixing by storage and/or by utilization. A large number of projects have been undertaken for proving techno-economics, reliability or safety of CO$_2$ storage. Large-scale demonstrations until 2010 are given in Table 1.

Table 1 : Active large-scale integrated CCS projects

Project name	Location/ Country	Type of industry	Storage option
Sleipner CO_2 injection	Norway	Gas processing	Deep saline formation
Snohvit CO_2 injection	Norway	Gas processing	Deep saline formation
In Salah CO_2 injection	Algeria	Gas processing	Deep saline formation
Weyburn-Midale CO_2 monitoring and storage	United States/ Canada	Synfuels Projection (pre-combustion capture)	EOR
Rangley Weber Sand Unit CO_2 injection	United States	Gas processing	EOR
Salt Creek	United States	Gas processing	EOR
Enid Fertilizer	United States	Fertilizer production (pre-combustion capture)	EOR
Sharon Ridge	United States	Gas processing	EOR

Source: Status of CCS, Global CCS Institute 2010 (see footnote 2)

A number of scientific, technical, economic, and regulatory barriers continue to be there in wider acceptance of the CCS technology for the power sector. A critique on CCS in UK by Haszeldine[4] optimistically suggests CO_2 storage as 'negative emission technology' and stresses on the political will and government support for it to succeed. The industry, however, needs cost-effective capturing of CO_2, scientific data for detail geo-modelling of reservoirs, understanding of cap rock structure and long-term continuous measurement, monitoring and verification (MMV) for the success of CO_2 storage experiments.

From these perceptions and developments, two important drivers emerge in CCS technology; one is about finding the possible technical solutions for CCS towards sustainable energy future and the second is about moving the focus of CCS research towards developing countries. Roman[5] while arguing that CCS is a political and strategic issue rather

[4] Haszeldine, Stuart. 2012. Carbon capture and storage, where is it? *Energy & Environment,* 23 (2 and 3):437–453

[5] Roman, M. 2011. Carbon capture and storage in developing countries: A comparison of Brazil, South Africa and India. *Global Environmental Change* 21:391–401

than simply a technological solution to the problem predicts that 91% of growth in global GHG emissions until 2030 is estimated to take place in the developing countries. Author then analyses situation for CCS deployment in Brazil, South Africa, and India. Among the three countries, India is seen as the most hesitant. It further adds that India has a relatively poor 'source and sink matching' and a long-distance pipeline through densely populated areas would be a major deterrent. From the policy viewpoint, the constraints are also geo-strategic border concerns with Pakistan and Bangladesh, as well as religious or social concerns for a suitable storage site, which is identified on the Indo-Gangetic plains close to the holy Varanasi city.

What would it mean to India? India is on the path of rapid economic growth. Although, power sector-installed capacity has steadily grown manifolds from 1300 MW in 1950 to 243,000 MW in March 2014, per capita consumption continues to be low. At present, India is struggling to meet basic energy needs and provide 'energy access' to all its population. It cannot afford to burden its people with almost double power tariff after application of CCS. As a climate change mitigation option, contribution of renewable sources in installed capacity is increasing and has become 30,000 MW. A target to reach a share of 20% in total by 2020 has been set. The share of clean energy technologies is 43% in the installed capacity (Figure 4). It comprised renewable (12%), hydropower (19%), nuclear (2%), natural gas and oil as clean fossil technologies (10%). Coal-based generation with 57% share continues as dominant energy resource for the next two to three decades. To sustain coal use, clean coal technology and CO_2 sequestration research are picking up in academic institutions and R&D centres. Voluntarily, India is committed to reduce Gross Domestic Product (GDP) intensity of CO_2 emissions by 20%–25% in 2020 over 2005 level, to be met entirely from demand sector by improving efficiency of products and processes. GHG emissions by sources and sinks in India in 2004 are summarized in Table 2. The GHG emissions from power and industry sectors together contributed to 80% of the emissions.

India's position on CCS in power sector can be summarized as follows:[6]

---

[6] Ministry of Power office memorandum dt. 24th April 2009

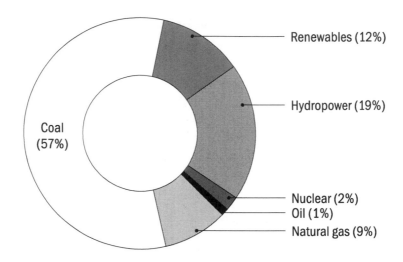

Figure 4   *Electricity installed capacity by resource in India in 2011*
Source: U.S. Energy Information Administration, International Energy Statistics.
India - Central Electricity Authority

Table 2 :   **Summary of GHG emissions in Giga gram (thousand tonnes) from India in 1994 by sources and sinks**

GHG – source and sink categories	CO_2 emissions	CO_2 removals	CH_4 emissions	N_2O emissions	CO_2 equivalent emissions*
All energy	679,470		2,896	11.4	743,820
Industrial processes	99,878		2	9	102,710
Agriculture			14,175	151	379,723
Land-use, land-use change and forestry	37,675	23,533	6.5	0.04	14,292
Waste			1,003	7	23,233
Total national emissions (Giga gram per year)	**817,023**	**23,533**	**18,083**	**178**	**1,228,540**

*Converted by using Global Warming Potential (GWP) indexed multipliers of 21 and 310 for converting CH_4 and N_2O respectively

Source: http://www.indiawaterportal.org/articles/greenhouse-gas-emissions-india-improvements-made-ghg-inventory-estimation

"India supports global efforts at R&D technologies aimed at reducing CO_2 emissions from coal-based industries. In this regard India supports R&D into CCS technologies which, at this time, still remain unproven. We have reservations as to its safety, cost, and performance of CO_2 storage and consequence of leakages. However, India will continue to participate in R&D activities and is willing to work on CCS technologies by deputing its scientists and engineers to sites in other countries where R&D into these technologies are being undertaken. We also support R&D into activities that seek to fix CO_2 to convert it into productive uses."

The 12th Five-Year Plan highlights on the role of CCS are in the context of new technologies like IGCC and energy efficient use of gas plants through facilitating distributed generation in CHP mode by developing network of gas pipelines must be encouraged. Developments in technology for CCS need to be carefully monitored to assess the suitability and cost effectiveness of this technology for Indian conditions.

Nevertheless, India faces several challenges in CCS development and clean energy research as the governance for, power, coal, petroleum and natural gas, renewable energy, atomic energy and heavy industry, responsible to the growth of energy sector, lies under various ministries, departments, and public sector units. Since fossil fuels use will continue to dominate for two to three decades, development of CCS may become important to address climate change concerns and further R&D is eminent. With this in view various laboratories under Council of Scientific and Industrial Research (CSIR), academic institutions and universities with the support from Ministry of Science and Technology and Ministry of Earth Sciences as well as several institutions under different ministries are currently engaged in CCS research. A nationally coordinated CO_2 sequestration research programme was conceived in 2006 to provide thrust to R&D under inter-sectoral committee in the Department of Science and Technology, Government of India.

Networking of these researches with the focus on knowledge sharing is required for finding solutions towards cost-effective and secured CCS technology. It is recommended that a 'National Commission on Energy Policy' would be the right choice to fulfil long pending agenda of integration of various energy ministries and address climate change issues under one umbrella. It has been recognized that innovative policies and role of technologies are important in the context of climate change and clean energy development. Such decisions will be taken at the highest level for the Commission would be responsible to address energy sustainability issues not only coal, power, renewable and nuclear energy, but also improved energy efficiency in demand sectors as well as new technologies and change-related related policy matters such as carbon capture, storage, and utilization (CCSU).

The present book is on CCSU. It covers various aspects of CO_2 capture, its utilization research and also takes a closer look at earth processes. The book has 17 chapters grouped into three sections as: Policy, Carbon Capture and Storage, and CO_2 Fixation and Utilization. The *first section* gives details of policy perspectives for India and an appraisal of policy and actions in coal-based economies. The *second section* is about CCS. It has five chapters dealing with capturing CO_2 and the underground storage options. CO_2 capture technologies can be broadly grouped into physical and chemical methods. Details of CO_2 storage in saline aquifers, minerals rocks, and abundant coal mines with research challenges are discussed. In the *third section* bio-sequestration methods for afforestation, biomimetic approach, use of CO_2 to ferrigate microalgae, production of value added products, chemicals and fuels are discussed. The CO_2 storage options for recovery of enhanced oil from depleted oilfields, enhanced gas recovery (EGR) from unmineable coal seams, and/or taking coastal protective measures such as artificial reef for ocean disposal are included. More detailed overview of what the individual chapters address is given below:

Chapter 1 on 'Carbon Capture and Storage: Can or Should We be so Indifferent?' is written by R.V. Shahi from Energy Infratech. CCS is among the advanced energy technologies suggested to make conventional fossil fuel sources environmentally sustainable. Tracing the history of India's participation in Carbon Sequestration Leadership Forum (CSLF), the author presents a policy dialogue on India's role towards this emerging technology as well as future perspectives. He suggests that it is to be looked as a possible long-term solution for climate change.

Chapter 2 on 'Policy and Regulatory Interventions in Abatement of CO_2 Footprints in India' is written by V.S. Verma, from Central Electricity Regulatory Commission. This chapter begins with evolution of power sector in India and transmission distribution challenges. The steps initiated with regard to CO_2 abatement and related regulations introduced from time to time are also discussed. In the total installed capacity of 250 GW, the share of renewable is approaching 30 GW. The technical issues related to renewable energy growth, its integration in grid transmission, and analysis of future capacity addition in the power sector towards a technology road map are also discussed.

Chapter 3 is on 'Carbon Capture, Storage, and Utilization – An Appraisal of Current Situation in Coal-Based Economies'. Written by Malti Goel and V. Johri of Jawaharlal Nehru University, Delhi, it presents the CCSU technology in technology sub-systems of multi-disciplinary education and research. Current coal use, CCS initiatives and regulations in selected coal-based economies namely the USA, India, PRC, Australia, and South Africa are reviewed. The USA is the world leader having the largest number of CCS projects as well as research publications. In South Africa, Centre for CCS has been actively pursuing the goal of CO_2 geological storage. Mapping of research output in these countries and initiation of capacity building activities in India are other highlights of this chapter.

The results of a 'CCS scoping study carried out in CO_2 emitting sectors' in the Indian context are presented in Chapter 4. The authors, Agneev Mukherjee, Arnab Bose, and Amit Kumar from The Energy and Resources Institute (TERI), Delhi, make an assessment of three major aspects of CCS – CO_2 capture, transport and storage, in both generation and demand sectors in India. A policy and legislation review is presented to highlight the barriers in CCS implementation. The need for further economic analysis and capacity development are the other issues raised in the chapter.

Chemical separation of CO_2 is a solvent absorption under low partial pressure. Physical methods make use of adsorptive separation technologies. A.K. Ghoshal, P. Saha, B.P. Mandal, S. Gumma, and R. Uppaluri from Indian Institute of Technology (IIT), Guwahati in Chapter 5 on 'Capturing CO_2 by Physical and Chemical Means', explain different absorption and adsorption processes for CO_2 separation/removal from flue gas. Post-combustion capture, absorbent synthesis and techno-economic feasibility studies are also discussed. Various pre-combustion technologies such as membrane reforming, IGCC are being tried for generating green power. Associated current and future CO_2 capture technologies for further research are described.

'Adsorption R&D for CO_2 Recovery from Power Plants' is covered in detail in Chapter 6. Written by A. Nanoti, M. O. Garg, A. Hanif, S. Dasgupta, S. Divekar, and Aarti from Indian Institute of Petroleum (IIP), Dehradun, it explains adsorption technologies that can be applied to both pre-combustion and post-combustion capture at high temperature and high pressure. A number of adsorbent materials such as zeolites, activated carbon as well as hydrotalcites for high temperature applications have been investigated in pre-combustion and post-combustion capture. Results from Pressure-Volume Swing Adsorption (PVSA) cycles for CO_2 recovery from flue gases and technology challenges in them are described.

Carbon dioxide captured from its point sources is to be stored away from the atmosphere. In little over a decade, geological storage of CO_2 has grown from a concept to a potentially mitigation option. Chapter 7 by A.K. Bhandari from Centre for Techno-Economic Mineral Policy Options (C-TEMPO) is on 'Geological Sequestration of CO_2 in Saline Aquifers'. It covers feasibility studies on saline aquifers, which are widespread and occur in almost all sedimentary basins. The author presents screening criteria for assessing the feasibility of storage in an aquifer. The advantages and constraints of CO_2 storage in saline aquifers, the studies undertaken to understand the mechanism of storage, and feasibility studies in India are discussed in this chapter. Further research would be needed to determine potential of the aquifers, particularly in view of lack of detailed knowledge available about them and absence of value addition expected from CO_2 storage in them.

Economic synergy from CO_2 storage can be expected from EOR of EGR. B. Kumar, formerly with National Geophysical Research Institute (NGRI), Hyderabad in Chapter 8 'Options for CO_2 Storage and Its Role in Reducing CO_2 Emissions' discusses options for CO_2 storage in these and basaltic rocks as well as technology advancements taking place such as in bio-carbon capture and geothermal storage. A brief overview of various storage options and increasing fertility of soil and ocean from CO_2 uptake are described. Some Indian initiatives are also covered in the chapter.

In Chapter 9 the assessment of 'CO_2 sequestration Potential for Indian Coalfields' is made by Ajay Kumar Singh and D. Mohanty from Central Institute of Mining and Fuel Research (CIMFR), Dhanbad. Coal in India occurs in two stratigraphic horizons as permian sediments in Gondwana basins in Peninsular India, and tertiary basins of the Eocene-Miocene age, mostly occurring in the north and north-eastern hilly regions. Information on coal and lignite deposits in India has been used to make a scientific assessment of CO_2 storage potential. Deeper unmineable coal seams are prospective candidates for CO_2 sequestration and also recovery of methane. This could form a basis for future policy about further research on identification of suitable sites for CO_2 storage.

Carbon management in forests is discussed in Chapter 10 in 'Carbon Sequestration Potential of the Forests of North-eastern India' by P. S. Yadava and A. Thokchom from Manipur University, Imphal. Forests play a significant role and act as sink for atmospheric CO_2, converting it into biomass and contributing to afforestation. North-eastern (NE) region of the country comprises seven states rich in flora and fauna. The forest cover is 66.07% of the total geographic area of the region. Carbon stock in NE states has been reported with data on carbon sequestration in the forest vegetation and in the soil. A unique opportunity exists for NE states to engage in mitigation of CO_2 because of high productive potential of the region.

In the *third section*, Chapter 11 deals with 'Prospects in Biomimetic Carbon Sequestration'. The authors are Shazia Faridi and T. Satyanarayana from University of Delhi (DU), South Campus. Carbonic anhydrases are fastest enzymes known for their efficiency in converting CO_2 into bicarbonates and then fuels. Microbes can thus act as efficient CO_2 sequestration media. They are natural, cost-effective, and a secure way of CO_2 storage and convert it into productive fuels such as methane and hydrogen. A policy direction to pursue biomimetic approach to carbon capture and combining with biochemical engineering can reap immense benefits.

In Chapter 12 on 'Bio-sequestration of Carbon Dioxide' by K. Uma Devi, G. Swapna, and K. Suman from Andhra University (AU), the potential of biological carbon fixation for remediation of CO_2 is explored. Flue gases from a power plant or industry having CO_2 as contaminant can be used to ferrigate the mass culture of microalgae and effectively sequester it. Opportunities and challenges in achieving success for large-scale application of this approach are analysed by the authors suggesting the need for further research on flocculants.

The Cyanobacterial Exopolysaccharides can utilize CO_2 and compensate for the changes in the earth's atmosphere by investing in a carbon concentrating mechanism (CCM). In Chapter 13 'Cyanobacteria in Carbon Dioxide Utilization and as Biosurfactants and Flocculants', R. Khangembam, M. Shamjetshabam, G. Oinam, A. Thadoi Devi, A.S. Sharma, P.P. Devi, and O.N. Tiwari from Microbial Bioprospecting Laboratory, Manipur have described the role of extracellular polymeric substances, which are complex in nature and can allow photosynthetic reduction of CO_2 to proceed efficiently. Further studies are, however, necessary to evaluate the feasibility of their application to mitigate CO_2 from the atmosphere.

V.K. Sethi in Chapter 14 discusses the efficacy of amine absorption system to separate CO_2 from the flue gases of a power plant and production of fuels such as hydrogen, methane, and biodiesel through algae route. Indian power sector is in transition from conventional power generation technologies to the green power technologies. A pilot plant comprising catalytic flash reduction of CO_2 using charcoal from gasifier, production of hydrogen from carbon monoxide, and production of methane using catalytic converter has been established at the Rajiv Gandhi Technological University (RGPV), Bhopal. The studies undertaken are described. Results of simulation studies for an operating thermal power plant are also presented.

The CO_2-EOR is a tertiary method for extraction of residual oil from the oil wells. A recent global study has indicated that over 50 large oil basins have reservoirs amenable for miscible CO_2 injection. In Chapter 15, Gautam Sen from Oil and Natural Gas Corporation on 'CO$_2$ Storage and Enhanced Oil Recovery' presents case studies from the USA and Canada and results of Weyburn–Midale large-scale project. Imaging of an oilfield using acoustic and elastic impedance measurements helps in monitoring of sub-surface flow associated with CO_2 injection, but there are challenges, which need to be addressed. Quantitative interpretation of three-dimensional seismic surveys would require further studies of reservoir temperature and pressure to compute CO_2 stored in the process.

The CO_2 injected in coal seams causes it to swell and due to higher adsorption preference of coal for CO_2 over methane, EGR can be expected. V. Vishal, T.N. Singh, S.P. Pradhan and P.G. Ranjith from IIT Bombay in Chapter 16 on 'Real-Term Implications of Carbon Sequestration in Coal Seams' analyse CO_2 sorption-induced coal matrix deformation, loss in permeability, and unconventional reservoir behaviour. Preliminary feasibility studies carried out in Raniganj coalfields in India suggest that coal seam CO_2 sequestration requires detailed reservoir characteristics and more site-specific studies.

Chapter 17, is on 'CO$_2$ Storage, Utilization Options, and Ocean Applications'. Written by M.A. Atmanand, S. Ramesh, S.V.S. Phani Kumar, G. Dharani, N. Thulasi Prasad, and S. Sadhu, from National Institute of Ocean Technology (NIOT). It deals with possible ecological and environmental concerns from direct disposal of excess CO_2 into the ocean. Indirect disposal and CO_2 utilization for developing artificial reef as protective coastal measures using industry wastes is a better option. Eco-friendly approaches to CO_2 sequestration in oceans are also described. Results of carbonation experiments carried out at the institute and pilot scale mass culture experiments using marine microalgae for production of biodiesel as well as challenges in them are discussed.

In conclusion the book is about communicating an emerging topic of environment science and technology with the students, researchers and policy makers. In it India's policy development studies and CO_2 capture/fixation research in developing new pathways are described. Technologies for CO_2 bio-sequestration and utilization, EOR from oil wells, and EGR from coal mines are discussed as possible climate change solutions for the coal-based energy industry. An overview of research frontiers as well as future policy perspectives is presented.

New Delhi

31 October 2014

Malti Goel

M. Sudhakar

R. V. Shahi

Contents

Foreword   *v*

Message   *vii*

Acknowledgement   *ix*

Introduction   **xi**

Section I: POLICY

1.   **Carbon Capture and Storage: Can or Should We be So Indifferent?**   **3**

R. V. Shahi

Introduction   3

Current Situation   4

Global Status of CO_2 Emission Trends   6

Cost of CCS Implementation   8

Future Role for India   9

Conclusion   11

References   12

2.   **Policy and Regulatory Interventions in Abatement of CO_2 Footprints in India**   **13**

V. S. Verma

Evolution of Power Sector in India   13

CO_2 Baseline Data for Power Sector   16

Steps Initiated by the Power Sector with Regard to CO_2 Abatement   18

Analysis of Future Demand and Capacity Addition in the Power Sector and Road Map for Technology Adaption   23

Conclusion   28

Summary   29

References   29

3. **Carbon Capture, Storage, and Utilization – an Appraisal
    of Current Situation in Coal-based Economies**   31

    Malti Goel and Vaibhav Johri

    Introduction   31
    Carbon Capture, Storage, and Utilization   32
    Current Scene in Selected Coal-based Economies   35
    CCSU Current Technology and Policy Status   39
    Mapping of CCS Research Publications   48
    Awareness and Capacity Building in CCS   51
    Conclusion   52
    Summary   53
    Acknowledgement   53
    References   54

4. **TERI'S Scoping Study on Carbon Capture and Storage
    in the Indian Context**   57

    Agneev Mukherjee, Arnab Bose, and Amit Kumar

    Introduction   57
    Country Background   58
    Economic Analysis   61
    Policy and Legislation Review   62
    Capacity Development Needs   65
    Conclusion and Recommendations   65
    Summary   66
    References   67

Section II: CARBON CAPTURE AND STORAGE

5. **Capturing CO_2 by Physical and Chemical Means**   71

    A. K. Ghoshal, P. Saha, B. P. Mandal, S. Gumma, and R. Uppaluri

    Introduction   71
    Post-combustion Carbon Capture   72
    CO_2 Capture by Adsorption   77
    Oxy-fuel Technology   80
    Chemical Looping Combustion   82

Pre-combustion Carbon Capture Methods   83
Integrated Gasification Combined Cycle   86
Gas Processing Prior to CO_2 Capture   87
Global R&D and Industrial Facilities for CO_2 Capture   88
Conclusion   89
Summary   90
References   90

6.   **CO_2 Recovery from Power Plants by Adsorption:**
    R & D and Technology   **95**

Anshu Nanoti, Madhukar O. Garg, Aamir Hanif, Soumen Dasgupta, Swapnil Divekar, and Aarti

Introduction   95
CO_2 Capture from Post-combustion Flue Gas   96
Application of Metal Organic Frameworks in CO_2
    Capture from Flue Gas   99
High Temperature CO_2 Capture Under Water
    Gas Shift Reactor Conditions   101
Conclusion   104
Summary   104
References   104

7.   **Geological Sequestration of CO_2 in Saline**
    Aquifers – an Indian Perspective   **107**

A. K. Bhandari

Introduction   107
Geological Storage of CO_2 108
Deep Saline Aquifers   109
Mechanisms of CO_2 Trapping in Saline Aquifers   110
Specific Issues Connected with Storage in Saline Aquifers   112
Screening Criteria for Storage in Saline Aquifers   113
Distribution of Saline Aquifers   115
Present Studies   119
Conclusion   120
Summary   121
Bibliography   121

8.   **Options for CO$_2$ Storage and its Role in Reducing CO$_2$ Emissions**   **123**

B. Kumar

Carbon Storage in Geological Formations   123
Innovative Carbon Storage Advances   127
Indian Initiatives and Recommendations   129
Summary   129
Bibliography   130

9.   **CO$_2$ Sequestration Potential of Indian Coalfields**   **133**

Ajay Kumar Singh and Debadutta Mohanty

Introduction   133
Coal and Lignite Deposits in India   134
Coal Inventory of India   135
Extraneous Coal Deposits of India   136
Grey Areas of Coalfields   137
Concealed Coalfields   139
Unmineable Coalbeds in Well-delineated Coalfields   139
Comparative Adsorption of CO$_2$ and CH$_4$ 140
Identification of Potential Coalbeds for CO$_2$ Storage   140
Conclusion   146
Summary   146
Acknowledgement   147
References   147

10. **Carbon Sequestration Potential of the Forests of North-eastern India**   **149**

P. S. Yadava and A. Thokchom

Introduction   149
Status of Forest Cover in North-eastern India   151
Growing Stock in Forest and Tree Out of Forest   152
Carbon Stock in North-eastern States   152
Rate of Carbon Sequestration   155
Carbon Sequestration in Soil   155
Soil CO$_2$ Emission Fluxes   157
Forest Management Practices in North-eastern India   158

Conclusion   159
Summary   160
References   160

Section III: CO$_2$ FIXATION AND UTILIZATION

11. Prospects in Biomimetic Carbon Sequestration   **167**

Shazia Faridi and T. Satyanarayana

Introduction   167
Bio-sequestration through Photosynthesis   168
Carbon Capture by Soil   168
Microbes in Carbon Fixation   168
Algal Sequestration of CO$_2$ 169
Heterotrophic Bacteria in Carbon Sequestration   170
Non-photosynthetic CO$_2$ Fixation by Heterotrophic Bacteria   170
Enzymatic Carbon Capture   171
Bio-mineralization using Carbonic Anhydrase   178
Enzymatic Capture of Carbon into Methanol   180
Conclusion   181
Summary   182
References   182

12. Bio-sequestration of CO$_2$ – Potential and Challenges   **189**

K. Uma Devi, G. Swapna, and K. Suman

Introduction   189
Mass Culture of Microalgae   191
CO$_2$ Mitigation from Flue Gas using Microalgae   192
Conclusion   200
Summary   201
References   201

**13. Cyanobacteria in Carbon Dioxide Utilization and as
Biosurfactants and Flocculants**   **207**

*Romi Khangembam, Minerva Shamjetshabam, Gunapati Oinam, Angom
Thadoi Devi, Aribam Subhalaxmi Sharma, Pukhrambam Premi Devi, and
O. N. Tiwari*

Introduction   207

Cyanobacteria in Carbon Capture and Utilization   208

Cyanobacterial Extracellular Polymeric Substances   213

Cyanobacterial EPS as Bioflocculants and Biosurfactants   215

Conclusion   217

Summary   218

References   218

14. **Production of Multi-purpose Fuels Through Carbon Capture and Sequestration**   **223**

V. K. Sethi

Introduction   223

Study Conducted at Rajiv Gandhi Technological University   225

Objectives of the Study   226

Methodology and System Configuration   226

Results of the Study   227

Conclusion   228

Summary   230

Acknowledgement   230

References   230

Appendix   232

15. **CO$_2$ Storage and Enhanced Oil Recovery**   **237**

Gautam Sen

Introduction   237

Case Studies from USA and Canada   239

Monitoring/Flow Surveillance   244

Conclusion   247

Summary   247

References   248

16. **Real-term Implications of Carbon Sequestration in Coal Seams**   **249**

V. Vishal, T. N. Singh, S. P. Pradhan, and P. G. Ranjith

Introduction   249

Conclusion   254

Summary   254

References   255

17. **CO_2 Storage, Utilization Options, and Ocean Applications** **257**

M. A. Atmanand, S. Ramesh, S. V. S. Phani Kumar, G. Dharani, N. Thulasi Prasad, and Sucheta Sadhu

Introduction	257
Major Sequestration Techniques	258
CO_2 Utilization in Ocean	259
Feasibility Study to use Converted Carbonates as Coastal Protection Structures	260
CO_2 Sequestration in Marine Microalgae	262
Conclusion	263
Summary	263
Acknowledgement	264
References	264

Index	*267*
About the Editors	*277*

Section I
POLICY

CHAPTER 1

Carbon Capture and Storage: Can or Should We be So Indifferent?

R. V. Shahi

CMD, Energy Infratech, New Delhi
and
Former Secretary, Ministry of Power
E-mail: rvshahi@energyinfratech.com

INTRODUCTION

India is a part of the United Nations Framework Convention on Climate Change (UNFCCC). The Kyoto Protocol to the UNFCCC was proposed as legally binding in 1997 and was ratified in February 2005. India acceded to the Protocol on 26 August 2002. The protocol targeted to achieve stabilization of greenhouse gas (GHG) concentrations in the atmosphere at a level that would prevent dangerous anthropogenic interference with the climate system [1]. It required Annex I countries to reduce their GHG emissions by an average of 5.2% below 1990 levels during its first commitment period (2008–12). Although India as a non-Annex I country does not have binding GHG mitigation commitments and was not given any target, Article 12.1 required each party to communicate a national inventory of GHGs, and a general description of steps taken for the implementation of the convention. India as non-Annex I has submitted First National Communication to the UNFCCC in 2004 and Second National Communication in 2012. The preparation of Third National Communication is in progress.

The USA, while not ratifying the Kyoto Protocol, launched a new initiative as Carbon Sequestration Leadership Forum (CSLF) in 2003 inviting both Annex I and non-Annex I countries as its members [2, 3]. India was also invited to become its member. In June 2003 in Washington, when CSLF Charter was being finalized, there was considerable discussion on "whether we should keep development and deployment of improved cost-effective technologies in the objectives of the charter?" The view of many developing countries was that deployment need not be included as it would be at least a couple of decades before a suitable and near cost-effective technology could be expected to emerge. After a lot of debate it was agreed not to include deployment and the following was finalized in its objectives:

"To facilitate the development of improved cost-effective technologies for the separation and capture of carbon dioxide for its transport and long-term safe storage; to make these technologies broadly available internationally; and to identify and address wider issues relating to carbon capture and storage. This could include promoting the appropriate technical, political, and regulatory environments for the development of such technology."

CURRENT SITUATION

Intergovernmental Panel on Climate Change (IPCC) assessments have particularly brought into focus the implications of ever-increasing carbon dioxide (CO_2) emissions. In spite of countries having committed to comply with the obligations under the Kyoto Protocol, the implementations have been far from being satisfactory. Instead of reducing the emission levels as compared to the base year 1990, in almost all cases emissions have increased in the countries which committed to reduce CO_2 emissions.

The initial phase of CSLF was for 10 years. However 10 years down the line, it is quite clear that it might take another couple of decades for any cost-effective technology to be fit for deployment to reduction of CO_2 concentrations. The number of CSLF projects has grown from 17 projects in 2007 having one project from India to 43 projects in 2013 with none from India. However, the CSLF Ministerial meeting, held in Beijing, China during 19–23 September 2011, decided to expand the scope of objective and agreed to extend and amend the CSLF Charter to include facilitation and deployment of technologies for utilization of captured CO_2. The author believes it was the failure on the part of the representatives of the developing countries in this meeting that this was agreed, as the situation is only marginally different today than in 2003, when the charter was finalized. In fact, no serious efforts were made during all these years to fund any major research project to develop cost-effective technologies. A sub-group, within the Policy Group, made recommendations in 2005 to finance major projects – funds to be contributed by member countries based on per capita emission, which could intensify the efforts on research and accelerate the pace of discovering cost-effective processes, suitable for ultimate deployment. This did not find favour from most of the representatives of developed nations and, therefore, it was not pursued.

It was only in the Ministerial Meeting, 4–7 November 2013 in Washington, that a reference was made about financing and incentive mechanism. "Recommended the development of financial frameworks and incentive mechanisms to drive near term demonstration and deployment of CCS and allow CCS technologies to compete with other low carbon technologies." Even this recommendation is of a very general nature requiring further follow-up on working out details of financing, sources of fund, and so on, which may again go into a long process. It suffers from being not specific and, therefore, lacks seriousness. As a matter of fact, the issue of finance had been discussed even in the CSLF meeting in September 2011 as reflected by the Communiqué issued on 22 September 2011. It says, "We today reaffirmed our commitment to work with private sector to build and finance the needed demonstration projects over the next decade." It took over two years for the Ministerial Meeting to endorse this by way of recommendation on financing, as mentioned above.

We are aware of the fate of implementation of the Kyoto Protocol. A lot was claimed by way of vision and commitment. But, a lukewarm approach by major players and highly bureaucratic systems and prosecutes made this wonderful concept deliver much less than what was expected. On the whole, the global response to carbon emission challenge has been more of a lip service and public relation exercise than a sincere and committed set of actions. The consequence has been obvious – increasing emissions rather than substantial reductions, which was the objective of the Kyoto Protocol. It is not to say, however, that nothing was done. In Europe the actions were significant but, on the whole, the outcome globally was disappointing. The CSLF initiative has been even more disappointing. The Zero Emission FutureGen Project was yet again a great idea. India was among the first few countries which contributed in 2005 to the fund to take it forward. The author was on the Steering Committee of this initiative. Right in 2006, in its meeting, there was more emphasis on formats rather than on substance. Unfortunately, this initiative has also not been driven with seriousness.

GLOBAL STATUS OF CO$_2$ EMISSION TRENDS

With a view to assess the ground realities in respect of global efforts made towards climate change mitigations, "Trends in Global CO$_2$ Emission: 2013 Report" of Netherlands Environmental Assessment Agency clearly brings out that Annex countries namely USA, EU, Germany, UK, and others, no doubt, reduced the per capita emission during the 20-year period (1990–2012), but the extent of reduction was much less than expected: USA–7%, Germany–23%, EU–19%, and UK–25% (Table 1). As in 2012, per capita emission in the USA was 16.4 tonnes, Germany 9.7 tonnes, UK 7.7 tonnes, Australia 18.8 tonnes, and Canada 16 tonnes. Maximum increase in per capita emission during 1990–2012 was in China (233%) at 7.1 tonnes up from 2.1 tonnes. Increase in India was from 0.8 to 1.6 tonnes, a rise of 100%. Obviously, in non-Annex I countries such as China, India, and others, which needed to develop their economies, increases were inevitable. However, both China and India have to remain engaged in evolving energy-efficient policies which would aim at furthering the growth requiring increased energy consumption, and, at the same time, adopt cost-effective technologies, which would not entail excessive CO$_2$ emissions. In this regard, the power policy of India which has incorporated deployment of Super Critical Technologies for coal-based power generation for major portion of its capacity addition programme is indeed a step in the right direction.

Moreover, the "Global Efforts on Carbon Capture and Storage Projects" report highlights that a number of projects have been undertaken in different countries, but very few of them are in coal-based thermal power generation plants. It would be relevant to quote from this report about the problems which arise out of excessive costs associated with CCS technologies.

Limitations of CCS for Power Stations

Critics say large-scale CCS deployment is unproven and decades away from being commercialized. They say that it is risky and expensive, and that a better option is renewable energy. Some environmental groups point out that CCS technology leaves behind dangerous waste material that has to be stored, just like nuclear power stations. Another limitation of CCS is its energy penalty. The technology is expected to use between 10% and 40% of the energy produced by a power station. Wide-scale adoption of CCS may erase efficiency gains in coal power plants of the last 50 years, and increase resource consumption by one third. Even taking the fuel penalty

Table 1 : CO$_2$ emissions in 2012 (million tonnes) and CO$_2$/capita emissions 1990–2012 (tonnes CO$_2$/person)

Country	Emiss-ions 2012	Per capita emissions					Change (1990–2012)	Change (1990–2012)	Change in CO$_2$ (1990–2012)	Change in population (1990–2012)
		1990	2000	2010	2011	2012				
Annex I*										
United States of America	5,200	19.6	20.6	17.6	17.1	16.4	−3.2	−17%	4%	25%
European Union	3,700	9.1	8.4	7.8	7.5	7.4	−1.7	−19%	−14%	7%
Germany	810	12.7	10.4	9.9	9.6	9.7	−2.9	−23%	−21%	3%
United Kingdom	490	10.3	9.2	8.2	7.5	7.7	−2.5	−25%	−17%	10%
Italy	390	7.5	8.1	6.9	6.7	6.3	−1.2	−16%	−9%	7%
France	370	6.9	6.9	6.2	5.8	5.8	−1.1	−15%	−5%	12%
Poland	320	8.2	7.5	8.7	8.7	8.4	0.2	3%	3%	0%
Spain	290	5.9	7.6	6.1	6.2	6.1	0.3	4%	26%	20%
Nether-lands	160	10.8	10.9	10.7	10.0	9.8	−1.0	−9%	0%	12%
Russian Federation	1,770	16.5	11.3	11.9	12.4	12.4	−4.1	−25%	−27%	−3%
Japan	1,320	9.5	10.2	9.7	9.8	10.4	0.9	9%	14%	4%
Canada	560	16.2	17.9	16.2	16.3	16.0	−0.2	−1%	24%	26%
Australia	430	16.0	18.5	19.4	19.4	18.8	2.7	17%	59%	35%
Ukraine	320	14.9	7.2	6.6	7.0	7.1	−7.8	−52%	−58%	−12%
Non-Annex I										
China	9,900	2.1	2.8	6.4	6.9	7.1	5.0	233%	293%	18%
India	1,970	0.8	1.0	1.5	1.5	1.6	0.8	110%	198%	42%
South Korea	640	5.9	9.8	12.2	12.9	13.0	7.1	121%	151%	14%
Indonesia	490	0.9	1.4	1.9	2.0	2.0	1.1	126%	213%	38%
Saudi Arabia	460	10.2	12.9	15.6	15.6	16.2	6.0	58%	177%	75%
Brazil	460	1.5	2.0	2.2	2.3	2.3	0.8	58%	109%	33%
Mexico	490	3.6	3.6	3.9	3.9	4.0	0.4	12%	58%	40%
Iran	410	3.6	5.2	5.2	5.3	5.3	1.7	47%	99%	40%
South Africa	330	7.3	6.9	6.4	6.3	6.3	−1.0	−14%	22%	42%
Taiwan	280	6.2	10.5	11.9	11.9	11.8	5.7	92%	121%	15%
Thailand	260	1.6	2.8	3.6	3.8	3.9	2.3	144%	189%	18%

Source of Population Data: UNPD, 2013 (WPP Rev. 2012)

*Annex I countries: Industrialized countries with annual reporting obligations under the UNFCCC and emission targets under the Kyoto Protocol. The USA signed but not ratified the protocol and thus the US emission target in the protocol has no legal status.

into account, however, overall levels of CO_2 abatement would remain high at approximately 80%–90%, compared to a plant without CCS. It is possible for CCS, when combined with biomass, to result in net negative emissions. Though, all of the currently (as of February 2011) operational BECCS (bio-energy with carbon capture and storage) plants operate on point emissions other than power stations, such as biofuel refineries. The use of CCS can reduce CO_2 emissions from the stacks of coal power plants by 85%–90% or more, but it has no effect on CO_2 emissions due to the mining and transport of coal. It will actually increase such emissions and of air pollutants per unit of net delivered power and will increase all ecological, land-use, air-pollution, and water-pollution impacts from coal mining, transport, and processing, because the CCS system requires 25% more energy, thus 25% more coal combustion, than a system without CCS.

Another concern is regarding the permanence of storage schemes. Opponents to CCS claim that safe and permanent storage of CO_2 cannot be guaranteed and that even very low leakage rates could undermine any climate mitigation effect. In 1986 a large leakage of naturally sequestered CO_2 rose from Lake Nyos in Cameroon and asphyxiated 1700 people. While the carbon had been sequestered naturally, some point to the event as evidence for the potentially catastrophic effects of sequestering carbon artificially.

On the one hand, Greenpeace claims that CCS could lead to a doubling of coal plant costs. It is also claimed by opponents to CCS that money spent on CCS will divert investments away from other solutions to climate change. On the other hand, CCS is pointed out as economically attractive in comparison to other forms of low carbon electricity generation and seen by the IPCC and others as a critical component for meeting mitigation targets such as 450 ppm and 350 ppm.

COST OF CCS IMPLEMENTATION

Although the processes involved in CCS have been demonstrated in other industrial applications, no commercial scale projects which can integrate these processes exist; the costs, therefore, are somewhat uncertain. Some recent credible estimates indicate that "the cost of capturing and storing CO_2 is USD 60 per tonne, corresponding to an increase in electricity prices of about US 6 cent (c) per kWh (based on typical coal-fired power plant emissions of 2.13 pounds CO_2 per kWh). This would double the typical US industrial electricity price (now at around 6c per kWh) and increase the typical retail residential electricity price by about 50%

(assuming 100% of power is from coal, which may not necessarily be the case, as this varies from state to state). Similar (approximate) price increases would likely be expected in coal-dependent country such as Australia, because the capture technology and chemistry, as well as the transport and injection costs from such power plants, would not, in an overall sense, vary significantly from country to country."

The CSLF has been issuing Technology Roadmaps. India was elected as Vice Chair to CSLF Technical Group in 2006. First roadmap came in 2004 and the latest in 2013. In the CSLF Technology Roadmap 2013, CO_2 utilization, particularly in the near term, is seen as a means of supporting the early deployment of CCS. First-generation CO_2 capture technology in 2013 has a high energy penalty and is in the demonstration phase. The second- and third-generation technologies for pre-combustion and post-combustion are discussing research, development, and demonstration needs as well challenges in industrial scale implementation (Table 2).

It can be seen that first-generation CO_2 capture technology for power generation applications are aiming to gain experience. Second- and third-generation CO_2 capture technologies are designed to reduce costs and the energy penalty. Possible targets for second- and third-generation CO_2 capture technology for power generation and industrial applications are 30% reduction towards 2030 and 50% reduction towards 2050 respectively, from first-generation levels in energy penalty, capital cost, and operation and maintenance (O&M) costs.

FUTURE ROLE FOR INDIA

India will need to expand its power generation capacity, over several decades, to ensure an accelerated economic growth. According to the Integrated Energy Policy (2006), by 2032, its capacity of power generation is projected to rise to 800 GW from the present of about 240 GW. Several measures have been taken to address the climate change concerns, including massive capacity addition through renewable energy generation – hydroelectric, wind, and solar. However, coal-based power generation will continue to occupy the dominant position. Therefore, it will be essential that the technological measures, which will aim at reduced consumption of coal on the one hand, and handling of CO_2 emitted, on the other hand, are adopted wherever such technologies are available. Indeed, while increased power consumption is a need for India, it is equally important that power is produced at a cost so as

Table 2 : CO$_2$ capture technology options of first-, second-, and third-generation and implementation challenges

Technology	CO$_2$ treatment first generation	Possible second- and third-generation technology options	Implementation challenges
IGCC with pre-combustion	Solvents and solid sorbents	Membrane separation of oxygen and syn-gas	Degree of integration of large IGCC plants versus flexibility
Decarbonization	Cryogenic air separation unit (ASU)	Turbines for hydrogen-rich gas with low NO$_X$	Operational availability with coal in base load Lack of commercial guarantees
Oxy-combustion	Cryogenic ASU Cryogenic purification of the CO$_2$ stream prior to compression Recycling of flue gas	New and more efficient air separation, example membranes Optimized boiler systems Oxy-combustion turbines Chemical looping combustion (CLC) reactor systems and oxygen carriers	Unit size and capacity combined with energy demand for ASU Peak temperatures versus flue-gas re-circulation NO$_X$ formation Optimization of overall compressor work (ASU and CO$_2$ purification unit (CPU) require compression work) Lack of commercial guarantees
Post-combustion capture	Separation of CO$_2$ from flue gas Chemical absorption or physical absorption (depending on CO$_2$ concentration)	New solvents (example amino acids) Second- and third-generation amines require less energy for regeneration Second- and third-generation process designs and equipment for new and conventional solvents Solid sorbent technologies Membrane technologies Hydrates Cryogenic technologies	Scale and integration of complete systems for flue gas cleaning Slippage of solvent to the surrounding air (possible health, safety and environmental (HS&E) issues) Carry-over of solvent into the CO$_2$ stream Flue gas contaminants Energy penalty Water balance (make-up water)

Source: CSLF Technology Roadmap 2013

to make it affordable for the people and also to keep the industry and manufacturing competitive with others.

Therefore, adoption and deployment of cost-effective technologies, to address the carbon emission concerns, becomes highly challenging. At the same time, perhaps we may not be right to expect that someone else would undertake research and development (R&D) to address our problems. India's position on CO_2 emission reduction, so far, has been that we continue to follow the policy of common but differentiated responsibility, and that highly developed nations, in which per capita emissions are many times more than the global average, need to effect drastic reductions, as also support technology development measures, has its own validity. However, having reached about 240,000 MW capacity to become 300,000 MW in the next two to three years and also considering the fact that a major portion of these additions would be coal based, it is time that India should think seriously in terms of substantial investments in R&D to adopt energy efficient technologies, and also carry out researches, on a long-term basis, to find its own solutions on CCS.

Obviously, these efforts will need to be networked and integrated with a number of such researches being carried out all over the world. Although, a number of academic institutions are currently engaged in research, the present approach of limiting these efforts to open-ended research has to change to collaborative work on research with targeted outcome. Organizations such as NTPC, BHEL, Coal India, ONGC, and large private sector power generating companies such as Tata Power, Reliance, and Adani Power, need to come together to formulate the required strategy. Part of National Clean Energy Fund may also be provided for such a programme. Progressive power companies are committed to allocate budget for R&D. Financial health of the large public sector companies, mentioned above, do provide enormous scope for funding such initiatives. Handling of CO_2 challenge, both pre- and post-combustion, needs more attention than has been, and is being given. We need to recognize that we cannot, and we should not, afford to remain so indifferent to this vital issue.

CONCLUSION

Global efforts in the last 50 years, but more seriously in the last 10 years, have achieved significant breakthrough in harnessing solar energy for power generation. In 2009, one could not believe that solar power price could come down to ₹6 to ₹7 per kWh in 2014 from ₹17 per kW initially. It is a reality today, and given the economy of scale, it can come down further. If similar efforts are not made for CCS, alternative strategy on economical renewable power would throw up a big challenge to power based on fossil fuels, even in terms of price of power. Therefore, coal-based power generation technology has two great challenges – one, efficiency of fuel has to be substantially higher from present 41%–42% in supercritical systems, and second, CO_2 emitted has to be managed effectively and economically. It may take a decade or two to reach a visible level of success, but it has to be done with serious initiative right now.

REFERENCES

1. Ministry of Environment & Forests. 2012. *Towards Preparation of India's Third National Communication and Biennial Update Report to UNFCCC.* New Delhi: Ministry of Environment & Forests, Government of India.

2. www.cslforum.org

3. Shahi, R. V. 2010. Carbon capture and storage technology: a possible long-term solution to climate change challenge. In *CO₂ Sequestration Technologies for Clean Energy,* S. Z. Qazim and Malti Goel (eds), pp. 13–22. Delhi: Daya Publishing House

4. http://www.cslforum.org/publications/documents/CSLF_Technology_Roadmap_2013.pdf

CHAPTER 2

Policy and Regulatory Interventions in Abatement of CO_2 Footprints in India

V. S. Verma

Former Member, Central Electricity Regulatory Commission
B-01, Swati Apartments, 12, I P Extension,
Patparganj, Delhi-110092
E-mail: vermavs2@gmail.com

EVOLUTION OF POWER SECTOR IN INDIA

In 1947, at the time of independence, the installed capacity of power generation in India was a meagre 1347 MW and the grid was fragmented. Now, as on June 2014, the installed capacity has grown close to 250,000 MW. The share of the private sector in it is about 30%. The gross generation has grown from about 4 BU in 1947 to 1000 BU in 2013–14. Per capita consumption of electricity was 16.3 kWh (units) in 1947 which has risen to 917.18 kWh in 2013–14. Sector-wise breakup of installed generation capacity, as on 31st August 2014, is shown in Table 1.

To transmit electricity by the high voltage supply system has now become more than 3 lakh kilometres of power grid, extending throughout the length and breadth of the country, one of the longest in the world. However, there are some gaps in the remotely located rural areas. The

Table 1 : Sector-wise breakup of installed capacity as on 31st August 2014

Sector	Installed capacity (MW)	% of total
Thermal (including gas and diesel)	176,118	69.5
Hydro	40,798	16.1
Nuclear	4,780	1.9
Renewable sources	31,692	12.5
Total	253,390	100%

Source: CEA

all-India transmission and distribution losses though reduced in the last decades, have remained high and are close to about 23%. During 2012–13 the peak demand met was 130,000 MW when the installed capacity was of about 200,000 MW. We were short in peak power by about 5%–6%. This was because part of the installed capacity consists of renewable energy sources (solar, small hydro, and wind) which do not contribute towards the real peak time load.

To reduce aggregate technical and commercial losses, in 2002–03 the government launched Accelerated Power Development and Reform Programme (APRDP). This was later restructured as R-APDRP by establishing baseline data in selected towns/cities, encouraging energy audits, and strengthening distribution networks. The Supervisory Control And Data Acquisition (SCADA) application and IT metering were also envisaged. Consequent to the Electricity Act 2003, decentralized distribution generation (DDG) under Rajiv Gandhi Grameen Vidyutikaran Yojna was launched in 2005 to provide all rural households in the country access to electricity. This was a landmark scheme for achieving electrification of rural households. An enabling framework was also proposed to introduce competition in the power sector through privatization of distribution networks.

As far as the power generating units are concerned, in the non-reheat range the unit size starts from 30, 60, 100 MW while in the reheat range it starts from 110, 120, 140, 200, 210, 250, 300/350, 500, 660, and 800 MW. The 660 MW, 800 MW, and some 1000 MW units (being conceived) have supercritical design parameters with an advantage in efficiency of about 1.5% to 2% over the subcritical ones that are in the range of 500–600 MW. The overall efficiencies of generation in thermal plants have increased over the years from a level of about 36% in pulverized plants to 40% in supercritical plants. There has been a technological upgradation of high order, while installing higher unit sizes, such as use of twisted turbine blades, to get higher levels of overall efficiencies.

The hydropower generating units have also seen some increase in unit sizes as well as efficiencies of operation. The largest unit sizes operating in the country on hydro side has been of about 250 MW. It may be pertinent to touch upon the planning options of the various sources of energy. While hydro is located in difficult geological terrain, nuclear fuel is not available in plenty, and the availability and price of gas is also very uncertain. Hence, for majority of power generation the only option is to depend on coal. Renewable sources in the form of solar photovoltaic (PV), wind, biomass are being inducted, but their costs are still on a higher side. The storage-type solar plants are also highly cost intensive. Indigenous researches are being conducted in this area to bring down the cost of generating power from solar and wind sources.

Transmission and Distribution

The systematic planning of power transmission in India started in the early 1960s. The country was then divided into five regional grids namely North, South, East, West, and Northeast. All the five grids used to operate independently at their own frequencies. In the late 1990s, the eastern regional grid comprising West Bengal, Bihar, Orissa, and Sikkim was running at high frequencies due to surplus generating capacity available in this region. The frequency used to touch 52–53 Hz and no commercial arrangements of penalties, etc. existed at that time. Consequently, over generation and under drawls were frequent and prominent.

In the early 1980s, interconnection of the five regional grids was planned. The inter-regional connections were designed in such a way that the capacity of these lines could meet only to the level of shortage in the concerned regions. The sector thus started growing in terms of installed capacity and demand for drawing electricity by the constituent members. A legally enforceable grid code was also evolved. Availability-based tariffs (ABT) were conceived and Central Electricity Regulatory Commission (CERC) spelt out appropriate regulatory measures through appropriate penalties for underdrawing and overdrawing of electricity from the grid. This commercial mechanism worked well over a period of time.

The experience and feedback available on the grid standards of Central Electricity Authority (CEA) and regulations of CERC with regard to grid operation and grid codes has been excellent. The commercial arrangements for penal rates for overdrawn and underdrawn of electricity during unfavourable conditions of grid operation have now become very deterrent. The operation bands of frequency range in the grid have now been tightened from 49.90 Hz to 50.05 Hz. This is almost in line with the international standards. The grid frequency levels of about 50 Hz have now been achieved in the system. The quality of electricity thus has become very acceptable. All the five grids in the country are now interconnected and the whole country is operating with one national grid. The responsibility of maintaining the grid disciplines lies with CERC.

CO_2 BASELINE DATA FOR POWER SECTOR

As discussed in the previous section, most of the generation in the grid (about 75%) comes from the coal-fired power plants. Accordingly, the emission of CO_2 (in absolute terms) is high and installation of clean coal technologies such as IGCC, adoption of supercritical steam parameters, and carbon capture and storage (CCS) have been considered as possible solutions. Ministry of Power, Government of India decided that while various options shall have to be evaluated and adopted to reduce CO_2 emissions, it was felt that India would not be in a position to support installation of CCS technology in its power stations for the time being due to various reasons. It is at this time the guidelines and action plans for reducing targets of CO_2 footprints by 2020 were evolved by adoption of various other alternatives. Table 2 gives the trends in CO_2 baseline data prepared by CEA from 2008 onwards. The CO_2 baseline data in the power sector was worked out by CEA in 2008 for the first time in the world with Indo-German association.

Table 2 : Trends in CO_2 baseline data for Indian power sector

S.No.	Item	2008–09	2009–10	2010–11	2011–12	2012–13
1.	Gross generation (GWh)	716,543	766,950	807,704	872,049	903,185
2.	Net generation (GWh)	668,029	714,680	753,106	813,306	838,485
3.	Share of hydro/nuclear (%)	18.7	17.1	18.4	19.6	16.9
4.	Absolute emissions (t CO_2)	548,593,836	580,075,201	597,724,599	637,261,050	696,303,676
5.	Combined margin (t CO_2/MWh)	0.86	0.90	0.91	0.95	0.98
6.	Weighted average emission rate (t CO_2/MWh)	0.82	0.81	0.79	0.78	0.83

Source: CEA

Adoption of CCS

Way back in the mid-1990s, the industrialized nations, who had already met most of their demands of energy by various sources including coal, hydro, and others, started making assertions that coal as the source of power generation resulted in the production of CO_2, the greenhouse gas (GHG) and such emissions must be brought down at any cost especially by the developing countries. The UN Framework Convention on Climate Change (UNFCCC) had been evoked and the reduction targets were on the agenda.

The pressure was more on India and China, the two developing countries, whose power demands were increasing at an exponential rate. However, the hard fact was that developed countries were required to marginally add the capacity to mainly replace the existing old units which have served beyond their useful life period. All these developed countries had fulfilled their requirement of power demands mainly from conventional resources of power such as coal, nuclear, hydro as well as natural gas. India and China were targeted mainly because the requirement of capacity addition in these countries seemed huge. The various governments thought it fit to impose direct or indirect impacts of their decision-making processes in these countries. This effect in all probabilities appeared to be driven by the manufacturers.

It was decided by the European Union (EU) that two experimental units for trial of the CCS technology shall be installed in India and China, and, in addition to these 10 more units shall be built in Europe and the USA. At that time there were series of discussions on this subject, both nationally and internationally. The details of CCS technology and its impact on efficiency, cost, etc. were however not shared. While studying the details of CCS technology for applications in the Indian context, it was discovered from the literature and data available in various publications that in the case of adoption of this technique for CO_2 abatement, the cost of generation of power almost doubles, the auxiliary power consumption increases, and the efficiency comes down by almost 30% (10%–12% points). The disaster management plans of such installations were also not spelt out. After a series of international deliberations, it became known that instead of 2 units, 10 units are to be installed in India and China and 2 units will allotted to European countries including Germany. The above decision-making process raised certain doubts on the seriousness of the proposals from EU countries and was not acceptable to India.

In view of the foregoing, attention was then focused by scientists in India towards conducting laboratory scale experimental research in relation to CCS technology. Further development of alternate technologies to utilize CO_2 separated from flue gases for useful purposes such as growth of algae, building materials, extracting of natural gases, etc. was also targeted. It is felt that research needs to be directed in the specific areas by breaking the problem into smaller components; for example one group has to take up sequestration of CO_2 from flue gases, other groups have to function on utilization of CO_2 for various applications such as conversion to useful construction materials. Similarly, yet another group may look at disaster management, economic, and other commercial aspects. The subject was kept alive with our scientists and technologists, and discussed in various national/international seminars/conferences.

STEPS INITIATED BY THE POWER SECTOR WITH REGARD TO CO_2 ABATEMENT

National Action Plan on Climate Change

Further as a matter of policy, it was felt that we should also prepare ourselves for meeting out any future eventualities as well as adopting measures to reduce CO_2 emissions. As a conscious decision of supporting the environmental aspects of reducing CO_2 emissions in the socio-economic sectors, National Action Plan on Climate Change was formulated by the Prime Minister's Office and released in 2008. Consequently, eight missions were set up that are as follows:

1. National Solar Mission
2. National Mission for Enhanced Energy Efficiency
3. National Mission on Sustainable Habitat
4. National Water Mission
5. National Mission for Sustaining the Himalayan Ecosystem
6. National Mission for a Green India
7. National Mission for Sustainable Agriculture
8. National Mission on Strategic Knowledge for Climate Change

Integrated multi-pronged long-term strategies, as proposed under each mission, are meant for achieving the set key goals in the context of climate change. While the National Solar Mission set the targets of solar capacity addition in the 12th, 13th and 14th plan (by 2027), the National

Mission on Enhanced Energy Efficiency (NMEEE) spelt out measures for increasing efficiency in industrial and power sectors in general by specific measures including technology upgradation as follows:

(i) Efforts are needed for increasing energy efficiency in generation, transmission, and distribution sectors. Energy efficiency in distribution would include reduction of transmission and distribution losses, demand side management to reduce the energy requirement, smart grid technologies, use of energy efficient equipment including home appliances and transformers, etc.

(ii) Energy efficiency needs to be introduced in manufacturing and production of goods by taking appropriate measures.

(iii) During the 11th Plan period (2007–12), it was estimated that around 11,000 MW of capacity addition has been avoided due to energy efficiency measures. The plan is to achieve around 12,000 MW avoided capacity during the 12th Plan through these measures.

Other Strategies in Power Sector

In the power sector to reduce CO$_2$ emissions, the following other strategies, covered in the National Action Plan Document, have been adopted:

(i) **Retire old inefficient plants** The inefficient and old units of coal-fired thermal plants should be retired, where there is no possibility of improvement in their efficiency of operation. About 15,000 MW such capacity consists of 30, 60, 100, 110, 120, 140 MW and some of the 200/210 MW were to be retired. Eventually, our average efficiency of generation would appear to be better. This would improve from the present level of about 34% to about 36%. Most of the units to be retired are non-reheat type with low inlet steam parameters. More than 5000 MW capacity is already retired. The CEA coordinates this activity.

(ii) **Renovation and modernization (R&M) of existing units** The older units having capacity of 200/210 and 500 MW ratings need to be renovated and modernized. This will bring back the efficiency of operation near to the rated values. New technologies should be adopted, such as improved blade profiles (twisted blades), etc. for steam turbines, which will increase the efficiency marginally beyond the original designed ones. These programmes are being taken-up on a case-by-case basis.

(iii)Adoption of supercritical technology   The new units to be adopted for future capacity addition in the upcoming plants shall have supercritical parameters that will give it marginally better efficiency (by 1.5%–2%) as compared to those with the existing subcritical 210/300/500/600 MW units. As per the capacity addition plans finalized by the government, all the capacity to be added in the 13th Plan shall be through supercritical units of larger ratings of 660/800/1000 MW. The 12th Plan capacity addition shall also see the majority of thermal units with supercritical parameters. About 25 units of 660/800 MW are now already working in the country on supercritical parameters. There are plans to introduce ultra-supercritical parameters on the future units to give an added advantage of further increase in efficiency.

Renewable Energy Development

In India, as in August 2014, the installed capacity for renewable energy reached 31.6 GW, that is 12.5% of the total electricity installed capacity. During the 12th Plan (2012–17), stress has been laid on further capacity addition of 33 GW through renewable sources of energy. This will include wind, solar, biomass, etc. However, there are issues with regard to application of these technologies; these are discussed here in brief.

- **Cost**   The costs of wind and solar power are rather on the higher side. The main reason for this is the monopolistic nature of supply of this technology. To bring down the costs of these technologies for adoption in India, scientific research institutes/IITs, educational institutions and the scientific community at large need to conduct research in this area which in turn needs sponsoring. Indigenization in the design and manufacture of all components of these technologies would definitely bring down the cost. Moreover, these have to compete on commercial basis with the conventional technology as far as the cost of generation is concerned. The question of cost is yet to be settled.

- **Grid connectivity**   Another issue which is bothering the power sector with regard to renewable sources of energy is grid connectivity and grid operation standards. The renewable energy generators ought to follow the grid connectivity standards and the grid code for operation and maintenance of these plants. The prediction of generation from wind and solar plants one day in advance at time intervals of

15 minutes during the day (96 time intervals) has to be made. As per the regulations brought out by CERC, the wind generations have been allowed a variation of ±30% on the predicted values of the generation on a one-day ahead basis, for each 15 minutes time intervals. The generators are also allowed to modify their schedules for four times in a day. But even this is not generally acceptable to the wind generators. A lot needs to be done in this regard. Meanwhile, some of the organizations have started complying with such requirements by adopting appropriate methods of prediction.

- **Peak load management and power tariff** The renewable energy sources by and large are not able to support on peak load management which normally occurs when these sources, especially solar, is not available. Hybrid models such as hydro-solar, hydro-wind, etc. would need to be considered for such operations. CERC has come up with attractive tariff structures for renewable sources of power generation and renewable energy certificate (REC) schemes to promote renewable energy in the country. These are covered in the following sections. The renewable energy purchase obligation (RPO) was considered an important tool in promoting renewable energy. To enforce RPOs in the states, cooperation of the respective State Electricity Regulatory Commissions (SERCs) and state governments is required. The RECs can be traded in the power market. Two such power markets, namely IEX and PXIL are operating in India today, one in Delhi and other in Mumbai. This is too vast a subject and could be deliberated separately. The CERC website could be consulted for the same.

Facilitating Measures for CO_2 Reduction

The following measures have been proposed by the Government, Ministry of Power (Bureau of Energy Efficiency, BEE) and CERC with regard to CO_2 reduction.

(i) **Renewable Energy Certificates** In order to facilitate full recovery of the cost of solar/wind/biomass or any other renewable sources of power and to resolve the issue of power tariff, the central commission designed a scheme called REC. 1 MWh of electricity generation from renewable sources of power entitles the generators to have one REC provided the power is sold at an average price of procurement of last year by the distribution company. The average price of procurement of power shall be declared by the distribution company each year. These RECs can then be traded in the energy markets at a price between the floor price and the maximum price notified by the central commission from time to time. This was to ensure the recovery of the price of renewable power by the generators. To make this operational, the Renewable Purchase Obligation (RPO) by the eligible entities has been notified by the respective SERCs. This, however, could not be enforced by the SERCs due to political and economic reasons and a large number of RECs are lying unsold. REC life (normally 365 days) is being extended by the central commission from time to time. The details can be seen on the CERC and Northern Regional Load Dispatch Centre (NRLDC) websites.

The concept of REC and RPO however needs a change. The author proposes developing a mechanism in which no one needs to buy REC. Instead it should only be an accounting procedure. The central load dispatch centre should be given the responsibility of booking from 3%–5% up to 10% as the case may be of power purchase to be deemed from the renewable source at the tariff decided by the regulatory commissions as applicable from state to state. The discretion of the distribution company to buy or not to buy due to cost consideration shall go away. In any case, the average impact of buying such power to the units of 5%–10% would have only marginal impacts (that is a few paisa only).

(ii) **PAT Scheme** NMEEE of National Climate Change Action Plan, BEE designed the Perform, Achieve and Trade (PAT) scheme. It is a market-based mechanism to enhance energy efficiency in the large energy-intensive industries. BEE has set the energy efficiency improvement targets for various industrial sectors and also for the power generating stations. These targets have been fixed based on the current operating parameters and the design efficiency parameters. The nine designated consumers under the scheme are depicted in Table 3.

The PAT Energy Efficiency Certificates thus awarded based on the achievements in performance improvement could also be traded in the power markets. RECs and PAT certificates are being made interchangeable. The scheme has come in operation with effect from 2012 and two years grace period was given. Accordingly, only after 2014–15, the assessment

Table 3 : Designated consumers for energy conservation under PAT scheme

S.No.	Sector	No. of identified DCs	Annual energy consumption (Mtoe)	Share consumption (%)	Apportioned energy reduction for PAT Cycle-I (Mtoe)
1.	Power (Thermal)	144	104.56	63.38%	3.211
2.	Iron and steel	67	25.32	15.35%	1.486
3.	Cement	85	15.01	9.10%	0.815
4.	Aluminium	10	7.71	4.67%	0.456
5.	Fertilizers	29	8.20	4.97%	0.478
6.	Pulp and paper	31	2.09	1.27%	0.119
7.	Textile	90	1.20	0.73%	0.066
8.	Chlor-Alkali	22	0.88	0.53%	0.054
9.	Total	478	164.97	100.00%	6.686

Source: [1]

year, the experience of operation of these schemes will be available. The main principle of operation of the scheme is 1 tonne of oil equivalent (toe) energy savings will give 1 PAT certificate. Better performance than the target fixed will give the PAT certificates equivalent to additional toe saved. Short fall in performance will call for a penalty of up to ₹ 100,000 as well as one will have to buy the PAT certificate from the market. The government will review the amount of penalties from time to time.

ANALYSIS OF FUTURE DEMAND AND CAPACITY ADDITION IN THE POWER SECTOR AND ROAD MAP FOR TECHNOLOGY ADAPTION

Table 4 gives the breakup of the future capacity addition during the 12th Plan (2012–17) – source wise and sector wise.

Renewable energy growth trajectory in terms of percentage contribution of total in terms of energy generation and total installed capacity is shown in Figure 1. Envisaged renewable energy capacity in different states is used as a benchmark (Table 5) to address the challenges of grid planning and grid operation in large-scale integration of renewable energy.

Table 4 : Capacity addition during 12th Plan as on March 2014

Sector	Hydro	Thermal breakup			Total thermal	Nuclear	Total	%
		Coal	Lignite	Gas/ LNG				
Central	6,004	13,800	250	827.6	14,878	5,300	26,182	29.57
State	1,608	12,210	0	1,712	13,922	0	15,530	17.54
Private	3,285	43,270	270	0.0	43,540	0	46,825	52.89
Total	10,897	69,280	520	2,539.6	72,340	5,300	88,537	
%	12.31				81.71	5.99		

Source: [2]

Note   Grid interactive renewable capacity addition likely during 12th Plan is about 30,000 MW (Wind-15,000 MW, Solar-10,000 MW, Others-5000 MW).

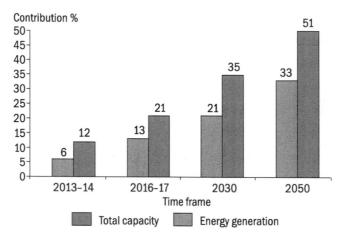

Figure 1   *Contribution of renewable energy in total capacity addition and in terms of generation*
Source: [3]

Major Challenges in Large-Scale RE Integration

Renewable energy generations may introduce new patterns in the power flows and a number of challenges exist in grid transmission. These can be identified as follows:

- Diurnal/seasonal generation variability (Figure 2).
- Wind/solar plants are generally located in remote locations and are away from load centre.
- Reactive power management in the grid.
- Low gestation period of 9–12 months for RE generation whereas development of transmission system takes 24–36 months.

Table 5 : Envisaged renewable capacity addition with state-wise breakup in RE-rich states in the 12th Plan

State	Existing capacity end of 11th Plan (MW)		Addition in 12th Plan (MW)		Total capacity (MW)	
	Wind	Solar	Wind	Solar	Wind	Solar
Tamil Nadu	6,370	7	4,339	3,014	10,709	3,021
Karnataka	1,783	6	3,619	253	5,402	259
Andhra Pradesh	230	22	3,150	1,677	3,380	1,699
Gujarat	2,600	600	3,368	1,361	5,968	1,961
Maharashtra	2,460	17	3,763	300	6,223	317
Rajasthan	2,100	200	2,181	3,513	4,281	3,713
J & K	-	2	12	102	12	104
Total	15,543	854	20,432	10,220	35,975	11,074
Total	16,397		30,652		47,049	

- Wind plants generally are not capable of supplying active power ramp up/down, reactive support to the grid in a despatchable and in controllable manner like conventional generators.
- In near bus fault situation, large wind generation may trip in certain pockets, magnifying the effect of electrical fault.

It is envisaged that green energy corridors of inter-state transmission systems in the renewable energy rich states of Andhra Pradesh, Gujarat, Himachal Pradesh, Maharashtra, Karnataka, Rajasthan, and Tamil Nadu will be built. In these states the wind patterns are variable on hourly basis, daily basis, and monthly basis. Wind variability in Rajasthan is shown in Figure 2. More emphasis is being given to real-time forecasts of renewable energy generation on various time intervals (15 minutes each, that is, 96 time intervals in a day) on the day ahead basis by increasing weather monitoring data stations. CERC has brought out very extensive regulations with regard to grid operation (grid code), which are really exhaustive for maintaining the grid discipline during the operation of power grid of mixed energy sources.

The analysis of anticipated total CO$_2$ emissions from the power sector at the end of the 12th and 13th Plans with the application of renewable energy system (RES) in different combination of energy sources is indicated in Figure 3.

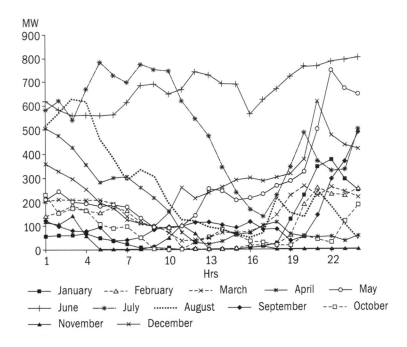

Figure 2   *Wind variability in Rajasthan*

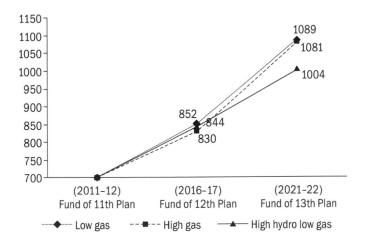

Figure 3   *Total CO_2 emissions (Mt) with RES [4]*

Future Road Map for Technology Adoption

Transmission systems need to be strengthened at intra-state as well as inter-state level following a large-scale increase in demand as well as steep increase in the number of generators in the system. There is a need to take integrated planning approach to address the concern of Dynamic reactive Compensation at strategic locations in the grid; Real-time dynamic state measurement scheme along with communication system; and Energy Storage Solutions in terms of large battery banks, compressed air system, etc. have to be found.

Renewable energy management centres

Renewable energy management centres need to be created and equipped with advanced weather monitoring systems and RE generation forecasting tools. Some of the balancing areas requiring strong grid interconnections are as follows:

* Forecasting of renewable generation.
* Availability of flexible generation (ancillary services), peaking generation, energy storage technology, etc., for supply/ demand balancing.
* Deployment of high-speed communication system such as fibre optics.
* Controllability of grid through establishment of Wide Area Management systems (WAMs) at RE pooling stations.

Smart grid

Smart grid allows demand side management, demand response through choice to the consumers, and energy conservation through consumer participation. It would also enable consumers to control the timing and choose the amount of power they consume based upon the prices of the power at a particular moment of time thus optimizing resources. Smart grid technologies would optimize the power supply situations and reduce the losses. Consequently, CO_2 emissions would get reduced due to integration of renewables and energy conservation measures. It will have the benefits of enhanced reliability of the grid by using real-time information.

Although initial steps for integration of smart grid were taken in R-APDRP, it includes adoption of IT applications for meter reading, billing and collection; energy accounting and auditing; generation of MIS; redressal of consumer grievances; establishment of data centre and IT-enabled consumer service centres, GIS mapping, automatic data logging and analysis, SCADA/DMS system, etc. From the communication technology viewpoint, it is necessary to have a uniform standard to allow interoperability and delivery of information among various devices connected to the grid.

Government has proposed to create 'India Smart Grid Task Force' to evolve the road map for implementation of smart grids. An 'India Smart Grid Forum' would help the Indian power sector to deploy smart grid technologies in an efficient, cost-effective, innovative, and scalable manner by bringing together all the key stakeholders and enabling technologies as a public-private-partnership (PPP) initiative.

CONCLUSION

To bring down the specific CO_2 emissions, greater use of renewable energy has to be promoted along with adoption of more efficient conventional coal-fired technologies such as supercritical technologies. However, the renewable sources of power, although encouraged throughout the world, have now been known to have created some serious problems of damaging the conventional machines of coal-fired power stations of larger size, such as 750-800-900 MW in the longer run. Power generator insulations may also break down, the blades/rotors may experience cracks, etc. For example Germany has installed 30,000 MW wind and 30,000 MW solar power plants. The power system in Germany caters to a total demand of 60,000 MW. The conventional plants capacity however, exceeds 120,000 MW. In a situation when wind blows and sun shines, the conventional power plants (coal-fired and lignite-fired) 750–800 MW units have to substantially back down generation to 20%–30% load. Such operations repeated frequently are known to cause damage to the machines.

Accordingly, while we address challenges of large-scale integration of RE through incentives and technology upgradation, it would also become necessary to make changes in the design features of conventional power plants in terms of using improved materials when constructing these plants and other related design features to cater to the load varying conditions. This would necessarily make the conventional generation more expensive.

Further considering the status of various technologies being adopted in the country as well as the actions initiated at the government level and strong regulatory interventions, it is concluded that Indian power sector would take progressive steps in meeting the demand with efficient generations, transmissions, and distribution systems. The CO_2 emissions shall also be contained within acceptable limits.

Summary

This chapter begins with the evolution of power sector in India and transmission and distribution challenges. The steps initiated with regard to CO$_2$ abatement and related regulations introduced for time to time are discussed. New initiatives and facilitating schemes such as renewable energy certificates for trading renewable energy capacity among states and PAT certificates for achieving CO$_2$ reduction though energy efficiency improvement by Central Electricity Regulatory Commission and Bureau of Energy Efficiency respectively, are described. An analysis of future capacity addition, challenges to be met in large-scale integration of renewable energy in the transmission sector, and future road map for technology adoption have been discussed.

REFERENCES

1. Chakravarty, K. K. 2014. *Current Status of Implementation of Perform Achieve and Trade (PAT) Scheme of India.* New Delhi: Bureau of Energy Efficiency

2. Verma, R. K. 2014. *Distribution.* A presentation in *T & D Conclave,* 13–14th May 2014, Mumbai from CEA (Personal Communication)

3. Prakash, Chandra. 2014. *Integration of Renewable Energy in State and National Grid.* A presentation in *T & D Conclave,* 13–14th May, Mumbai from CEA (Personal Communication)

4. Verma, V. S. 2010. CO$_2$ mitigation: issues and strategies. In *CO$_2$ Sequestration Technology for Clean Energy,* S. Z. Qasim and Malti Goel (eds), pp. 171–181. New Delhi: Daya Publishing House

Carbon Capture, Storage, and Utilization – An Appraisal of Current Situation in Coal-based Economies

Malti Goel[1] and Vaibhav Johri[2]
[1]Former Scientist 'G', Department of Science and Technology
and
CSIR Emeritus, Scientist
[2]Ex-Project Scientist, Centre for Studies in Science Policy,
Jawaharlal Nehru University,
New Delhi-110067
E-mail: maltigoel2008@gmail.com

INTRODUCTION

For long, capturing carbon dioxide (CO_2) has been shrouded in mystery. It emerged as a solution to climate change mitigation as an end-of-pipe solution to control global pollution in the atmosphere from fossil fuel-based energy systems. Significant strides are being made in industrializing carbon capture, storage, and utilization (CCSU) – a CO_2 sequestration technology. CO_2 sequestration involves capturing of excess CO_2 from its point sources and its permanent fixation through storage or utilization away from the atmosphere. The various methods of CO_2 capture are basically derived from gas separation techniques, which include chemical absorption, membrane separation, physical adsorption, and cryogenic separation at the source [1–2]. Captured CO_2 is then sequestered by means of surface processes or by sub-surface storage and/or by utilization in recovery of energy fuels and minerals [3]. If the source and the underground fixation sites are not near to each other, transport of CO_2 in liquid form over longer distances is required. Utilization of captured CO_2 into value-added products has been given importance [4].

Science and technology gaps exist in cost-effective CO_2 capture processes and materials as well as in the development of site-specific models for its fixation. In this chapter, we introduce CCSU technology in its seven sub-systems of multi-disciplinary research and education. Current coal use in carbon capture and storage (CCS) in selected coal-based economies – the USA, Australia, PRC (People's Republic of China), South Africa and India – CO_2 emissions, climate change actions, and a review of policy perspectives are discussed. Mapping of CCS research output during 1990–2011 in these countries has been carried out in two epochs of 11 years each and is presented. Lastly, objectives and highlights of capacity building workshops on CCS conducted in India are stated.

CARBON CAPTURE, STORAGE, AND UTILIZATION

The CCSU technologies can be grouped into seven major technology sub-sets as given in the ensuing sections.

Post-Combustion Capture

Post-combustion capture involves chemical, physical, and cryogenic methods for capturing excess CO_2 from its point sources such as flue gas in thermal power plants. Chemical separation is a solvent-based process involving absorption and desorption. Amine solvent-based chemical separation process for CO_2 separation has been widely accepted, but not yet adopted on a large-scale as it is cost intensive [5–6]. In the cost-distribution curve of CCS (utilization not included), the capture cost has the highest share. Research is needed in the development of new types of solvents that can become cost-effective for large-scale application in operational plants. Physical separation is the development of materials and techniques for adsorption and filtration, which include ceramic materials and membranes as filters. The challenges are in materials science research and pressure swing adsorption techniques. Cryogenic separation option is based on cooling of various components in the flue gas. It however, is a highly energy-intensive process and not in practice at present.

In-Combustion Processes

Supercritical and ultra-supercritical combustion technologies increase efficiency of power generation process. The concentration of CO_2 in flue gas is higher and it facilitates reduction in the cost of capture. These along with alternative coal combustion techniques, viz. oxy-fuel generation and chemical looping are proposed as in-combustion CO_2 mitigation options. While supercritical technology is being implemented in more than 400 plants the world over, there is no CO_2 capture on plant scale yet. Research is directed towards development of materials for ultra-supercritical boilers and reduction in the cost of oxygen separation from air, which are major challenges for development of oxy-fuel combustion technologies.

Pre-Combustion Technology

Pre-combustion capture has been proposed for capturing CO_2 from gasified or liquefied coal prior to combustion. In this, coal is converted into gas or liquid and before it is put in energy cycle. For capture of CO_2 from coal gas, a mixture of carbon monoxide and hydrogen processes such as hydrogen membrane reforming, shift gas reaction in association with Integrated Gasification Combined Cycle and Fischer-tropch synthesis are adopted. Coal-to-Liquid (CTL) and other coal conversion technologies facilitate pre-combustion CO_2 capture at high temperature and high pressure.

CO_2 Transportation and Injection Technologies

The captured CO_2 can be transported to appropriate storage sites in either gas, liquid, or solid form. For gas transportation compression is a must, otherwise at the atmospheric pressure it would occupy huge volume. For long-distance transportation liquid form of CO_2 is preferred, wherein it can be transported through existing oil pipelines and/or tankers on trucks. Knowledge about 'Geostratiphic Acceptance' is required to ascertain the mode of transportation [7]. If the distance between the source and the sink is more, the share of pipeline infrastructure cost in the total cost can increase by about 15%–20%. For marine transportation, temporary storage facility at the ports is required. In the solid form, CO_2 is transported as CO_2 ice.

Transportation is followed by underground injection in the appropriate geo-physical environment with the help of mining engineers. The injection technologies are dependent on geometry of wellhead(s), injection wells, cap rock structure, and geomorphology of the region.

CO_2 Underground Geological Storage

A number of underground or sub-surface geological storage options for CO_2 are proposed. Indepth studies of different geo-physical environments viz., saline waters/aquifers, empty oil and gas fields, mineral rocks, coalfields, basalt rocks, shale reservoirs as well as undersea gas hydrates have been undertaken by researchers across the world. Among others, basalt formations are considered as attractive storage media as they are ancient hot lava sites. However, CO_2 storage mechanism in basalts is not fully understood yet [8]. The saline aquifers are other promising storage sites. The Sleipner demonstration project, first to begun in Norway, is injecting 1 million tonnes (Mt) of CO_2 per year since 1996 in the North Sea. It is providing proof-of-concept for reliability and safety in saline aquifers storage. In the mineral rocks, as CO_2 plume migrates, some of it may react with formation minerals to precipitate as carbonates [9]. Permanent CO_2 storage in deep sea sediments/gas hydrates are being explored by marine scientists as a possibility.

In the sub-surface, CO_2 is stored in the supercritical phase, which is attained at a temperature of 30.41 K and pressure of 73.8 bars. Various trapping mechanisms under study include physical or stratigraphic trapping, mineralogical trapping, geochemical mixing, and residual gas mixing. The US Department of Energy has classified 11 major types of reservoirs for knowledge development pathways of CO_2 sub-surface sequestration [10], each requiring considerable research on its performance for long-term storage and safety. Detail geo-modelling studies of different reservoirs; cap rock structure; and measurement, monitoring and verification (MMV) techniques are important components for securing underground storage of CO_2 [11–12].

CO_2 Fixation in the Terrestrial Ecosystem

Carbon assimilation occurs in forests, trees, crops and soil, and these act as CO_2 sinks. Terrestrial sequestration of CO_2 is chiefly biological or surface bio-sequestration process. Both captured CO_2 and CO_2 concentrated flue gas can be utilized. Besides photosynthesis fixation in plants, concentration of CO_2 in the atmosphere is also influenced by soil carbon pool and local environmental factors [13–14]. Advanced biomimetic approach using immobilized carbonic anhydrase is becoming important in bio-reactors and can have significant CO_2 sequestration potential. Mining waste, heavy metal industry slag, and marine algae sequestration are other effective terrestrial fixation options. Cultivation of marine algae for conversion into biodiesel and process optimization can have added benefits. Accelerating the natural weathering process of CO_2 absorption and enhanced carbonation mineralization of olivine/silicate rocks offer potentially low-cost solution for CO_2 sequestration [15]. The carbonation reaction is exothermic and products are thermodynamically stable.

CO_2 Utilization Technologies

Utilization of captured CO_2 makes it an attractive proposition as a risk-free option and results in value added products. CO_2 utilization follows chemical sequestration routes on the surface, sub-surface, and in the oceans. In the surface utilization, CO_2 can be converted into production of ethanol or methanol, fertilizers, feedstock in food processing and carbonated drinks, and so on. As such, CO_2 has low chemical activity, but it is possible to activate it towards chemical reaction by application of temperature or pressure or by use of appropriate catalysts. In a bio-reacting medium such as microalgae it can be converted into fuels, pharmaceuticals, and value added products. In the sub-surface, CO_2 utilization envisages injection in depleted oilfields or in un-mineable coal seams for producing enhanced oil or enhanced coalbed methane (ECBM), respectively. It can provide an economic synergy to CO_2 sequestration process. Innovative CO_2 disposal by producing carbonization products has been proposed in the oceans.

In addition to these, extraction of pure carbon from captured CO_2 can add huge value to the utilization process.

CURRENT SCENE IN SELECTED COAL-BASED ECONOMIES

Coal Resources

Globally, coal resources are estimated at 861 billion tonnes (Bt). The global coal consumption, which at present is contributing to 68% of the increase in CO_2 emissions, has been growing (inset in Figure 1). Global CO_2 emissions have become 38 giga tonnes (Gt) in 2012. Coal being an important energy resource has grown at a faster rate in the Asia-Pacific region. The region accounts for 67% of global production in 2011 as compared to 27% in 1981. It is projected to take a share of 71.3% of the total global consumption in 2015 and 77.7% in 2030, with PRC and India as largest consumers [16]. A study of five coal-based economies taking into account the contribution of coal to be about 50% or more in the total electricity capacity and initiatives taken for carbon sequestration research has been made. Table 1 has coal production data for PRC, the USA, India, Australia, and South Africa in 2006 and 2012 as well as

Table 1 : Increasing coal production and its current share in total energy for selected countries

Country	Coal production in 2006 in Mt	Coal production in 2012 in Mt	Share in total energy % in 2010
PRC	2,482.0	3,650.0	81
USA	990.0	935.0	50
India	427.0	595.0	68
Australia	309.0	431.2	69
South Africa	244.0	260.0	94

Source: World Coal Institute, 2012

its percentage share in total energy. The total global production of coal increased from 4.4 Bt in 2006 to 5.8 Bt in 2012. Coal production was highest for PRC followed by the USA.

CO_2 Emissions and Targets

The growth of global CO_2 emissions from 1870 onwards is depicted in Figure 1. According to International Energy Agency (IEA) analysis in 2012, global coal-fired power plants accounted for more than 8.5 Gt of CO_2 emissions per year.

International Energy Agency database on coal-fired plants in them and the total coal-based installed capacity in 2010 is presented in Table 2 [17]. Coal-fired plants in PRC are 2928, higher in number and capacity

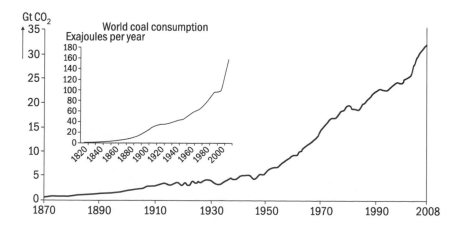

Figure 1   *Trends in coal consumption and CO_2 emissions from 1870 onwards*
Source: CDIAC, Oak Ridge National Laboratory, USA

Table 2 : Coal-fired plants in 2010 [17]

Country	No. of plants	Total installed capacity in GW
PRC	2,928	669.2
USA	1,368	336.3
India	809	100.5
South Africa	114	37.5
Australia	109	29.97

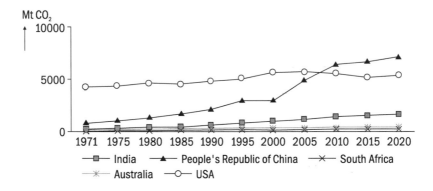

Figure 2 *Trends in CO_2 emissions in five selected countries from 1971–2010*
Source: CDIAC

than remaining four countries combined together. The country-wise CO_2 emissions trends from 1971 onwards are depicted in Figure 2.

Due to anthropogenic economic activities, the Intergovernmental Panel on Climate Change (IPCC) has estimated an observed global average temperature increase by 0.74°C during 20th century [18]. Using various Global Circulation Models for atmospheric circulation in earth system, it is predicted that the average global temperature could increase between 1.1–6.4°C by 2100. In a global drive towards reduction of greenhouse gas (GHG) concentrations in the atmosphere, almost all countries of the world are part of international conventions and protocols on climate change. National actions for climate change mitigation have been initiated by several countries. In Table 3, we present a summary of gross domestic product (GDP) per capita, CO_2 emission reduction commitments, and date of ratification of international climate change protocols for the selected economies in the descending order of their GDP per capita.

Table 3 : CO_2 emission reduction commitment of selected countries

Country	GDP per capita in USD (2011)	Emission reduction target and pledges				Date of ratification	
		By 2020 (Unconditional)	By 2020 (Conditional)	By 2050 (Conditional)	Other	UNFCCC	Kyoto Protocol
Australia	67,039	–5% relative to 2000	Up to –15% or – 25% relative to 2000	–80% relative to 2000		30 Dec 1992	12 Dec 2007
USA	49,922	–17% by 2020 relative to 2005		Towards a goal of – 83% relative to 2005	–30% in 2025, –42% in 2030 relative to 2005	15 Oct 1992	–
South Africa	8,090		–34% relative to BAU		–42% by 2025 relative to 2005 and missions to peak between 2020 and 2025	29 Aug 1997	31 Jul 2002
PRC	5,439		–40 to –45% of CO_2 per unit of GDP to 2005		–17% of CO_2 per unit of GDP by 2015 relative to 2005	5 Jan 1993	30 Aug 2002
India	1,528		–20% to –5% emission intensity per unit of GDP relative to 2005			1 Nov 1993	26 Aug 2002

Source: Compiled using data from various sources including http://unstats.un.org

CCSU CURRENT TECHNOLOGY AND POLICY STATUS

Among the different options for CO_2 mitigation in the atmosphere, CCSU is receiving attention as an emerging energy technology from both scientific community and policy makers. Technology sub-sets of CCSU described in the earlier section have been researched for long, but their integrated or application as an approach to CO_2 reduction is new. The technology is infrastructure intensive and current status is being assessed in terms of policy initiatives and number of research and demonstration projects. The USA took lead by launching multi-country Carbon Sequestration Leadership Forum (CSLF), an initiative of Department of Energy, USA in 2003 that has focus on collaborative R&D. Australia has come forward to establish Global Carbon Capture and Storage Institute (GCCSI) in 2009 aiming to accelerate commercial development of CCS technology.

According to Global CCS Institute, a total of 74 CCS projects in 2012 were in various stages of planning, development, demonstration, or operation; of these only eight large-scale projects are in operation. The CO_2 emissions per capita, policy perspectives, and the level of activity in CCSU in different countries are shown in Figure 3. In all the five countries viz., Australia, the USA, PRC, Australia, South Africa, and India, an R&D policy exists and international collaborations are taking place. Information and market-based instruments and regulations and standards are mostly present in Australia and the USA.

In the following section, we describe current situation in terms of status of CCS projects, climate change policies, and a framework for policy in respect of CCS large-scale implementation in the selected economies in the descending order of GDP per capita (in 2011).

Australia

Australia is rich in natural resources like petroleum, natural gas as well as coal. It is has the fourth largest reserves of coal of about 76,400 Mt and was fourth largest producer of coal until (in) 2011. About 70%–75% of the electricity in Australia is produced from coal. In 2010, the total emissions according to the National Greenhouse Gas Inventory were 560.77328 Mt from energy, industrial processes, agriculture, waste and land use (land use change and forestry) sectors. With growing energy demand, it is observed that CO_2 emissions have increased by 166% during 1971–2010. Although, Australia did not ratify Kyoto Protocol until

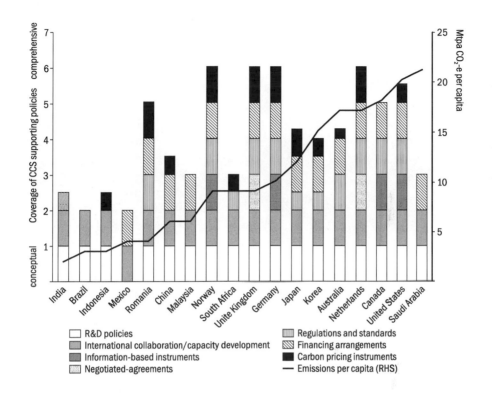

Figure 3   *Level of CO₂ emissions and CCS supporting policies*
Source: GCCSI [19]

2007, it has taken a lead in establishing GCCSI. Australian research focus has been on large-scale demonstration plants and it has 21 CCS projects in various stages of development and implementation. The first collaborative project Cooperative Research Centre for Greenhouse Gas Technologies (CO₂CRC) to demonstrate CO₂ storage in Otway basin began test injection in 2008 with continuous MMV. Gorgon project planned as commercial-scale demonstration, launched in 2009 for injection of 120 Mt CO₂ in saline aquifers is another large-scale project. Location of major, active and planned CCS projects in Australia is depicted in Figure 4. Australia has a National CO₂ Infrastructure Plan to study potentially suitable sites to store captured CO₂ and speed up the development of transport infrastructure near major CO₂ emission sources.

Australia's major initiatives for the development and deployment of CCS are as follows:

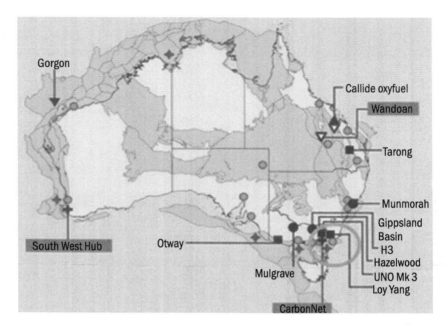

Figure 4   *Australia's major active and planned CCS projects*
Source: CO$_2$CRC

(i)   Establishment of a not-for-profit GCCSI in April, 2009. The institute is expected to build and share the expertise on CCS globally and to effectively implement the CO$_2$ emission reduction strategies through CCS collaborative projects. The GCCSI has taken up the task of information dissemination through publications and webinars.

(ii)   Initiation of the CCS flagship programme, which includes research as well as demonstration projects. A total amount of $1.9 bn (in 2012) has been earmarked for construction of two to four commercial scale CCS projects with coal-based plants of 1000 MW capacity over a period of nine years.

(iii)   Introduction of carbon tax on coal.

While Australia is taking an active role in carbon management and promotion of CCS projects, there is also considerable on-going debate on the proposed reduction targets and the potential role of CCS in reducing GHG emissions [20].

The United States of America

The USA is the largest economy of the world with a GDP of approximately USD 15.09 trillion in 2012. It has the world's third largest population

(313 million) and is the second largest emitter of CO_2 after PRC. Nearly one-third of the anthropogenic CO_2 emissions of the USA come from power plants. The CO_2 emissions from coal combustion (almost entirely used for electric power production) increased over 18% between 1990 and 2003 with a forecasted 54% increase by 2030 in business-as-usual scenario. However, with new findings of shale gas and its increasing production, the CO_2 emissions have actually reduced by 1.4% in 2012. Although, it is not bound by Kyoto targets (having not ratified it), about 7.5% reduction of emissions has taken place in the USA between 2006 and 2011, which according to IEA is more than any other country. Under the Copenhagen Accord, the USA has set an emissions reduction target for 2020 in the range of 17% against a 2005 baseline.

The USA has been a leading country in the world in CCSU R&D and demonstration projects. Carbon capture and sequestration research in programme mode began in 1989 in the Energy Laboratory of Massachutes Institute of Technology (MIT) [21]. An industrial consortium on Carbon Sequestration Initiative was launched in 2000. International CSLF started in 2003 has 23 developed and developing countries as its members. The CSLF can be stated as part of its science diplomacy of the USA and took the lead to promote collaborative R&D. A 'Charter' was devised based on consensus among the members. The CCS activities in the USA also include creation and maintenance of seven Regional Carbon Sequestration Partnerships (RCSPs) throughout the country.

According to National Energy Technology Laboratory (NETL) database [22], there are more than 240 CCS projects across the world in various stages of initiation, planning, demonstration, or operation stages including those abandoned. Among these, the USA is having the largest number of 113 projects including 24 large-scale projects. The USA allotted $3.4 bn to clean coal and CCS technology including industrial CO_2 capture projects. FutureGen, Weyburn, Frio, Mountaineer, and Alison pilot projects are in progress in gas-fired/coal-fired plants for demonstration of CO_2 capture (both pre-combustion and post-combustion) and fuel recovery. The CO_2 utilization is also receiving attention.

The Department of Energy, USA has been working for developing a legal framework for CCS and for its large-scale deployment. These include CCS related bills introduced on (i) Carbon Capture and Sequestration Programme Amendments, (ii) CO_2 Capture Technology Prize Act in 2011, (iii) Rule governance about CO_2 geological sequestration and Underground Injection Control Programme stated by Environment

Protection Agency (EPA). In addition to these, EPA has also developed output-based performance standard for CO_2 per megawatt hour in new fossil fuel-fired power plants. Several regulatory issues are being resolved with respect to CO_2 storage such as those dealing with pore space ownership, long-term liability, etc.

South Africa

South Africa has the world's ninth largest coal reserves of about 30,156 Mt and sixth largest coal producer in the world in 2011. Of the total electricity generated, 93% comes from coal. Its location at the natural point of convergence to both Atlantic and Pacific markets makes it is also one of the largest coal exporter. South Africa is the world leader in coal liquefaction technologies and much of its energy needs is being fulfilled by the coal-based liquid fuels. The dependence on coal resulted in increased CO_2 emissions from industrial utilities and power generation facilities. According to the World Resources Institute, with its 400 Mt of CO_2 emissions annually, South Africa is rated among the world's top 20 emitters. The amount of CO_2 has increased by 121% since 1971, but the reduction commitments are also high. South Africa has adopted energy strategy compatible with Copenhagen Accord to reduce emissions by 34% by 2020 and 42% by 2025 below business-as-usual (Table 3).

In South Africa, National Environmental Management Act (NEMA) has been imposed for taking measures for pollution control. These provisions may apply for CCS projects, which form part of South Africa's energy and climate change policies. Having ratified the UNFCCC in August 1997, it acceded to the Kyoto Protocol in 2003. The Clean Development Mechanism (CDM) and other climate related funding opportunities such as clean coal technology development are being explored. To identify promising directions for research in CCS, a road map for CCS has been prepared after joining CSLF in 2004. A high level assessment of CCS potential in South Africa has been carried out. The South African Department of Energy has also developed a CCS legal and regulatory framework in 2011. An entry into the IEA, Legal and Regulatory Review was submitted. One of the main recommendations of the 17th Conference of Parties (COP–17) meeting hosted in Durban in 2011 under UNFCCC [23] has been inclusion of CCS in CDM.

A strategy for introducing carbon tax, application of CCS for coal-fired power stations, and not approving new coal-fired stations without carbon capture readiness exists. In 2008, a CCS ready requirement was placed on the Kusile power plant. South Africa has come forward to establish a centre for CCS in the South African National Energy Research Institute (SANERI) in 2009. The centre is private/international/public partnership funded by local industry, government, research institutes, and international agencies. A National Climate Change Response Green Paper was prepared in 2010 and National Climate Change Response Policy has been stated.

The South African Centre for CCS (SACCCS) has been actively pursuing the goal of carbon geological storage. An atlas based on a literature review of all available boreholes and other geological information which identified and ranked potential CO_2 storage sites mostly located in Mesozoic basins along the coast and further to the Karoo basin has been prepared. It is estimated that approximately 60% of the total emissions of CO_2 annually can be captured and be potentially available for sequestration [24]. It is aimed to test-inject CO_2 into underground storage cavities and expected to store 10,000–50,000 tonnes of CO_2 from 2017 onwards and to install a CCS demonstration plant by 2020.

People's Republic of China

People's Republic of China is the world's most populous country. It is also the fastest growing economy. PRC has the third largest reserves of the coal of 114,500 Bt, which accounts for 14% of the world resources. It became the largest coal producing country in the world in 2011 and with this a vast expansion of coal-based power plants has taken place. In the installed capacity of 700 GW (2011), about 79% of the electricity generated is coal based. The electricity capacity is projected to double by 2020. A dramatic growth has taken place in the last decade; PRC has overtaken the USA and become the world's largest CO_2 emitter (although per capita emissions are still lower than many advanced countries). The amount of CO_2 emissions has risen by 800% since 1971. The CO_2 reduction targets for PRC are 40%–45% of CO_2 equivalent per unit of GDP in 2020 against 2005 baseline as depicted in Table 3.

People's Republic of China has the highest number of CCS projects in the Asia-Pacific region. A total of 11 projects are in various stages of development. With the experience and confidence gained from implementing the pilot projects, a number of Large-Scale Integrated Projects (LSIP) are being proposed. As of May 2012, the GCCSI recorded six LSIPs in PRC and this number is expected to increase further [25]. The LSIPs in PRC have pursued CO_2 capture technology such as pre-combustion capture and oxy-fuel combustion technologies. In carbon storage technologies, there is a strong focus on utilizing the captured CO_2 for commercial applications that may generate revenue. The captured CO_2 for enhanced oil recovery (EOR) is in practice since 2006 and a majority of the proposed CCS projects have considered EOR as the preferred choice of carbon storage.

In June 2007, Government of PRC in its report on "China's Scientific and Technological Actions on Climate Change" devised a road map and the need for capacity building for research, development, and demonstration of CCS projects was established. It also reduced carbon intensity (CO_2 emissions per unit of GDP) by 17%; reducing energy intensity (energy consumption per unit of GDP) by 16% and increasing the share of non-fossil energy to 11.4% [26]. In PRC, Ministry of Science and Technology has issued S&T action plan on climate change-related scientific research in conjunction with other ministries. The CCS Technology Policy has been prepared. Under its key technologies R&D programme, strategic studies on CCS relating to applicability of CCS and its potential impacts on energy system in mitigating GHG emissions have been launched.

The CCS demonstration GreenGen project aims at coal gasification-based hydrogen power generation and CO_2 storage technologies development as a means to near zero emission coal-based power plant. China Huaneng Group Corporation was initiated to launch a CO_2 capture demonstration in a 250 MW IGCC plant, Quashai. PRC is also one of the first Asian countries to initiate a pilot project on ECBM recovery by CO_2 sequestration in coal beds. SINOPEC, a Chinese company is currently operating an integrated pilot plant that captures 0.04 million tonnes per annum (Mtpa) of CO_2 for EOR. It is also planning to expand the capacity of this facility up to 1 Mtpa CO_2 capture in Phase II, which is expected to be completed in 2014. CO_2 injection for ECBM in coal mines has been tested. Shenhua Company started feasibility studies of CCS in a CTL plant and is proposing to store captured CO_2 in the saline aquifer in Ordos basin. Two CCU pilot projects are currently in operation which sell food-grade CO_2 for food and beverage production.

On the policy front, China's National Development and Reform Commission has issued a notification in 2011, which requires all new coal-chemical demonstration projects to be capable of reducing CO_2 emissions. PRC became a member of CSLF in 2003. It has active international cooperation in CCS projects with EU (European Union), UK (United Kingdom) and Canada, and has launched NZEC-China-UK Near Zero Emissions Coal Initiative and COACH (Co-operation Action within CCS China) - EU. Asian Development Bank is also helping PRC to reduce its CO_2 emissions through CCS [27].

India

India is a rapidly growing economy. Being a rich country in mineral and human resources, it is leading in information and software technologies. It is the third largest consumer and producer of coal in the world, of which about 75% of coal is consumed in generating electricity. Currently, 57% of electricity installed capacity comes from coal-based thermal power. In the total installed capacity of 207.9 GW (December, 2012), coal-based capacity is 118.7 GW, which has a share of 69% in generation. Capacity addition of 36 GW from ultra-mega power projects is in pipeline. According to Geological Survey of India (GSI), India is having a total coal resources of 293.5 Bt and reserves proven of 118 Bt. Coal will continue to be a dominant resource for meeting the basic needs of electricity. In the 12th Five-Year Plan, it is projected to add 78 GW coal-based power generation and further 100 GW in the 13th Five-Year Plan using highly efficient supercritical and ultra-supercritical coal combustion technologies.

India's early policies and programmes have been directed towards energy conservation and adaptation to climate change. In terms of technology development, the major priority since the 1970s have been increasing efficiency of generation by adopting best practices in thermal power generation and as well as demand side management by conservation of resources and introducing renewable energy. Centre for Power Efficiency and Environmental Protection (CENPEEP) was created by the National Thermal Power Corporation (NTPC) in 1994. National Mission on Enhanced Energy Efficiency (NMEEE) is providing thrust to schemes like PAT (Performance, Achieve and Trade) and has set targets for efficiency improvement in nine industry sectors. Department of Science and Technology, Government of India with the aim to accelerate research towards controlling GHG emission supported joint technology projects on energy efficiency improvement in the end-use sectors of economy [28].

Policy measures towards adaptation to climate change are enumerated by Ministry of Environment and Forests. The CO_2 emissions as per India Greenhouse Gas Inventory 2007 are 1497 Mt [29]. Being a non-Annex I country, India is not bound by Kyoto commitments, but has voluntarily set CO_2 reduction targets of 20%–25% in 2020 relative to 2005 baseline. National Action Plan on Climate Change was formally launched by Prime Minister's Council in 2008 during the 11th Five-Year Plan with its eight core areas to be pursued in the mission mode for adaptation and mitigation of climate change. While dependence on coal-based generation is high, India is emerging in renewable energy sector with the launch of Jawaharlal Nehru National Solar Mission in 2010 and is having 12.5% share of renewable energy at present in the total installed capacity.

In accordance with a policy for abatement of climate change, government support to carbon capture and fixation research is considered appropriate as one of the potential tools to limit GHG emissions. At the national level, National Programme on CO_2 Sequestration research commenced in the Ministry of Science and Technology in 2006–07 and the following thrust areas were identified [30]:

(i)   Modelling of terrestrial agro-forestry sequestration,

(ii)   Carbon capture process development,

(iii) Bio-fixation of microalgae and CO_2 sequestration, and

(iv) Policy development studies.

In 2007, Indian Carbon Dioxide Sequestration Applied Research (ICOSAR) was launched to facilitate information sharing on developments, activities, policy studies with stakeholders. India's CCS specific research, policy, and gap areas have been discussed by Shackley and Verma [31], Shahi [32], and Goel [33].

At the international level, India is founder member of CSLF since 2003 and became Vice-Chair to CSLF Technical Group in 2006. India is a member to many international co-operations that work for the development and dissemination of CCS technologies. These include joining USA in its FutureGen project; Big Sky Carbon Sequestration partnership; the Asia Pacific Partnership for Clean Development and Climate. On individual basis, new laboratory collaborations started with National Energy Technology Laboratory, USA; Pacific Northwest National Laboratory, USA and SINTEF, Norway. India is one of the founding members of GCCSI and the premier utility NTPC has joined as industry representative. A National Clean Energy Fund (NCEF) has been set up in 2010 for funding research and innovative projects in clean energy technology [34], which plans to undertake a technological mission to develop carbon sequestration technologies.

The deployment of CCS is not India's policy and, therefore, currently there are no legislations concerning it. There are three or four major concerns about CCSU technology; first, the technology is not yet developed to commercial application, second, high investment of the order of ₹ 100 billion per annum in CCS is required for its adoption, third, CCS would add enormously to the price of electricity, which would make it unaffordable to the large population of the country, and the fourth, there is the vital question of availability of suitable resources and associated risk/safety issues in underground storage. There are many other specific concerns about application of technology [35]. Economic policy studies of 'CCS in India' and possible linkages of 'CCS with CDM' have been carried out. Existing legislation that may govern CCS industrial activities in future are discussed [36]. Presently, no large-scale projects are being planned in India except for one by Indian Farmers Fertilizer Limited, which is in operation using amine technology for CO_2 capture.

MAPPING OF CCS RESEARCH PUBLICATIONS

In the light of the above policy scenario, we make an assessment of current level of CCS research in the five countries. Among the various bibliometric and bibliographic parameters available for such an assessment, research publications analysis can be considered a systematic record of the research output. It is said to be a reflection of research investment at a place [37]. It is also one of the means to determine future growth in technology by new processes and products. Hence, we analysed R&D output in CCS in terms of number of scientific papers on a year-to-year basis. Measurement of R&D output in the five selected economies has been carried out using database of 'Web of Science', which is the collection of the online journals and academic citation index provided by Thomson Reuter. The data was searched from 1990–2011 by the keywords, which were found during the literature review of the subject. The keywords ("post-combustion", "pre-combustion", "oxy-fuel", "carbon capture", "carbon storage", "CO_2 capture", "CO_2 separation", "CO_2 removal") were arranged in the Boolean format to get the most accurate searches for the field. The database from Science Citation Index was considered.

The results showed a total of 1422 papers in all years including articles proceeding papers, editorial materials, news items, meeting abstracts, reviews, book chapters as well as letters. Of these proceeding papers (166), reviews (95) and articles (1136) were selected and refined by subject-wise categories of related disciplines. The results were now 1297 papers. The year-wise research output is shown in Figure 5. It is seen that there is a phenomenal growth in the research output from 2006 onwards and in 2011 the output resulted in 267 papers and exceeded 300 in 2012.

Further, the research output is broken into two time periods, viz., in the last decade (1991–2000) and recent developments (2001–11). During 1991–2000, the total research publications were 117 with the highest output in a year being 20 research publications, which occurred in 1993. Comparison of data for selected five countries, in two epochs of 11 years each 1900–2000 and 2001–11 is shown in Figure 6. The research output has grown more than eight times during 2001–11 to that in 1991–2000. Looking at the country-wise research output, the USA has the highest output with a total of 273 research papers. Australia and PRC occupy second and third positions, respectively, and there were 86 for Australia and 80 for PRC research publications during the period of study. In both, the number of publications for 'Web of Science' during

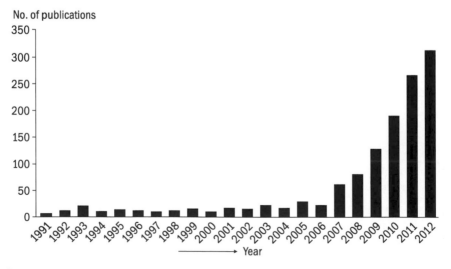

Figure 5   *Growth in research publications in CCS from 1991–2012*

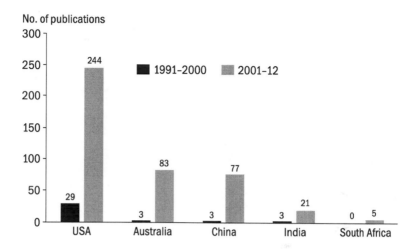

Figure 6   *Comparison of research publication in CCSU for selected countries during 1991–2000 and 2001–2011*

1991–2000 was only three and rose to 83 for Australia and 77 for PRC during 2001–11. The research output in database of 'Web of Science' for India is 26 as against 24 in Elsevier's Scopus for same set of keywords. South Africa also has less number of research papers in CCS, although it had considerable more output in clean coal technology.

Discussions and Future Steps

A trend analysis is presented tracking the developments taking place in CO_2 capture, storage, and utilization research. From 2005 to 2012, global CCS research publications in numbers have increased almost ten times. During this period, research has been catalysed and new research networks have been created [34]. Results of technology-wise search showed highest number of publications in post-combustion followed by oxy-fuel research. Pre-combustion research received lowest priority. In post-combustion research, absorption has the highest number of research papers aimed at cost reduction. The research output in carbon storage in various underground reservoirs is significant; however, due to regional variations results from one region cannot be applied to another. Saline aquifers provide largest potential storage capacity, but result did not show many publications. The results for value addition research from sub-surface storage that is, utilization of captured CO_2 in EOR, enhanced gas recovery (EGR), and ECBM recovery are shown in Figure 7. From South

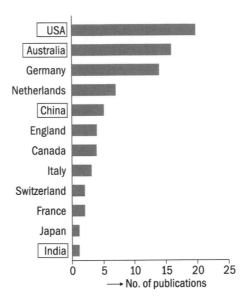

Figure 7   *Number of publications for value addition CCS (EGR, EOR, and ECBM)*

Africa, no result was seen in the period of study. Additional technology specific data are needed to fill in the gaps.

A major issue in CCSU penetration has been pilot and demonstration projects. Big collaborative projects such as CO_2 GeoNet (a consortium of 13 EU countries), CASTOR, CO_2 SINK, Snohvit, In Salah and others are implemented by EU. This analysis has a bias on coal-based economies. EU is not included in the study. Despite these limitations, it gives a fair representation of status. Further research would be benefitted from inclusion of databases such as Scopus, which has a wider coverage of scientific journals. More detailed analysis of journal impact factors and linkages with projects is recommended.

AWARENESS AND CAPACITY BUILDING IN CCS

International Energy Agency CCS Roadmap envisions 20 demonstration projects by 2015, 100 CCS projects globally by 2020, and 3000 projects by 2050. The CCS is expected to become commercially viable only in 2030 [38]. The science of CO_2 sequestration technology involves basic physics of global warming as well as chemical and biological effects of CO_2 fixation in the earth system. Need for capacity building in CCS is thus highlighted in many forums such as CSLF, GCCSI, IIEA-GHG, and other multi-lateral organizations [39].

India has been giving thrust to CO_2 sequestration research and capacity building needs. Two five-day Capacity Building Workshops have been held as ACBCCS-2009 and ACBCCS-2013. ACBCCS is an acronym for Awareness and Capacity Building in CCS. ACBCCS–2009 was organized in Delhi from 27–31 July, 2009 at Indian National Science Academy. It originated from the concern for environment and need for knowledge sharing among different CCS groups [40, 41]. The workshop was unique, held for the first time in India for industry stakeholders. It was acclaimed as a very good and timely initiative, structured to create awareness and dissemination of recent developments in CCS.

The second five-day national level workshop ACBCCS-2013 was held from 15–19 January, 2013 at India International Centre, New Delhi to provide an exposure to enhanced world view and broaden the perspectives for Indian research community. The technical papers of the workshop addressed a range of topics in Carbon Capture and Storage: Earth Processes. It provided a platform to young researchers from different academic institutions across the country to learn more about carbon capture, utilization and storage related earth processes, its relevance, barriers, and opportunities.

The themes covered in these capacity building workshops include carbon capture and sequestration, pre-combustion, combustion and post-combustion options, bio-sequestration, terrestrial sequestration, earth processes utilization, and storage in the oceans. Integrated approach in terms of R&D has been deliberated. R&D being one of the means by which one can expect future growth. In the current scenario as India gears to achieve a global presence in energy industry, it is the right time to institutionalize the research networking in CCSU.

CONCLUSION

The CCSU comprises inter-disciplinary, infrastructure-intensive technologies. CCSU technology sub-systems are dealing with various capture, transport, fixation, storage, and utilization options. Technologies can be grouped in seven sub-sets. Technology learning curves suggest that each technology sub-set for commercialization is facing many challenges and opportunities. Starting from CO_2 capture, which requires cost-effective, energy and material saving technologies to be developed, challenges exist in CO_2 compression, transportation, and fixation. Storage technologies need proof-of-concept. Permanence of storage and no safety risks are the main criteria for sub-surface storage.

The current situation of CO_2 emissions, CCS projects, and policy in five selected coal-based economies of the USA, India, PRC, Australia, and South Africa is reviewed. Measurement of R&D output in the five selected economies is carried out using database of `Web of Science'. USA has been the world leader in a number of projects as well as research publications. Australia and PRC are the second highest, while Australia is taking an active role in promotion of CCS projects, PRC is in the process of leading Asia in LSIPs and in policy development. South Africa is giving thrust to overcome barriers for CCS through regulations. India has a focus on CO_2 fixation research and is growing in research output. Several government agencies and industries have taken initiatives to promote and perform CCSU research. Capacity building efforts are being made. It is concluded that more country specific studies are needed to assess CO_2 sequestration potential and economic feasibility of using emerging technologies such as CCSU for climate change mitigation.

Summary

This chapter dealt with the emerging topic of CO_2 sequestration, which is among the advanced energy technologies. Various processes for CO_2 capture, utilization, and recycling in the earth system were described. Termed as CCSU, it is being discussed in national and international forums. The CCSU research has been in focus for the past two decades. An appraisal of current situation in five coal-based economies of the world viz., the USA, Australia, PRC, South Africa, and India in terms of coal use, penetration of CCS technologies, climate change actions, and policy infrastructures to promote CCS were presented. As the number of CCS projects is growing steadily, research output is also growing. Mapping of research output in the selected economies is carried out. The objectives of capacity building workshops held in India were discussed.

ACKNOWLEDGEMENT

Malti Goel acknowledges with thanks the Emeritus Scientist grant received from HRDG, Council for Scientific and Industrial Research, New Delhi - vide 21.0759.2009 for carrying out this study.

REFERENCES

1. Herzog, H. J., E. M. Drake, and E. E. Adams. 1997. CO_2 capture, reuse and storage technologies for mitigating global climate change: A White Paper, Final Report. *Energy Laboratory, Massachusetts, Institute of Technology*

2. Metz, Bert, Ogunlade Davidson, Heleen de Connick, Manuela Loos, and Leo Meyer (eds). 2005. *IPCC Special Report on Carbon Capture and Storage, Working Group III.* Cambridge, UK: Cambridge University Press

3. Bachu, S., W. D. Gunter, and E. H. Perkins. 1994. Aquifer disposal of CO_2: Hydrodynamic and mineral trapping. *Energy Conversion Management* 35:269–279

4. Parsons Brinckerhoff and Global CCS Institute. 2011. Accelerating the uptake of CCS: Industrial use of captured carbon dioxide. *Global CCS Institute,* March

5. Herzog, H. 2009. Carbon dioxide capture and storage. In *The Economics and Politics of Climate Change,* Dieter Helm and Cameron Hepburn (eds), pp. 263–83. Oxford: Oxford University Press

6. Rao, A. B. and E. S. Rubin. 2002. A technical, economic and environmental assessment of amine-based CO_2 capture technology for power plant greenhouse gas control. *Environ. Sci. Technol.* 36:4467–4475

7. McGrail, B Peter, E. C. Sullivan, and M. D. While. 2010. Adapting integrated sequestration system design for densely and not so densely populated areas. *34th Course of the International School of Geophysics* 25–30 September, Erice, 25–30

8. Prasad, P. S. R., D. Sarma, L. Sudhakar, U. Basavaraju, R. S. Singh, Z. Begum, K. B. Archana, C. D. Chavan, and S. N. Charan. 2009. Geological sequestration of carbon dioxide in Deccan Basalts: Preliminary laboratory study. *Current Science* 996:288–291

9. Goel, Malti, S. N. Charan, and A. K. Bhandari. 2008. CO_2 sequestration: Recent Indian research. In *IUGS Indian Report of INSA 2004–2008,* A. K. Singhvi, A. Bhattacharya, and S. Guha (eds), pp.56–60. New Delhi: Platinum Jubilee Publication

10. Buchanan, R. and T. R. Carr. 2011. Geologic sequestration of carbon dioxide in Kansas. *Kansas Geological Survey, Public Information Circular (PIC)*27

11. Bickle, M., A. Chadwick, H. Huppert, M. Hallworth, and S. Lyle. 2007. Modelling carbon dioxide accumulation at Sleipner: Implications for underground carbon storage. *Earth and Planetary Science Letters* 255(1–2):164–176

12. Goel, Malti. 2012. Sustainable energy through carbon capture and storage: Role of geo-modeling studies. *Energy & Environment* 23:299–317

13. Wisniewski, Joc, R. K. Dixon, J. D. Kinswan, R. N. Sampson, and A. E. Lugo. 1993. Carbon dioxide sequestration in terrestrial ecosystem. *Climate Research* 3:1–5

14. Yadava, P. S. 2010. CO_2 mitigation: Issues and strategies. In *CO_2 Sequestration Technologies for Clean Energy Future*, S. Z. Qasim and Malti Goel (eds), pp. 163–170. Delhi: Daya Publishing House

15. Schuling, R. D. and P. Kigjsman. 2006. Enhanced weathering: An effective and cheap tool to sequester CO_2. *Climate Change* 74:349–354

16. Umbach, Frank. 2011. The future of coal, clean coal technologies and CCS in the EU and Central East European countries: strategic challenges and perspective. *EUCERS Strategy Paper No. 2*, 12 December

17. International Energy Agency. 2012. *World Energy Outlook.* Paris: International Energy Agency

18. Intergovernmental Panel for Climate Change (IPCC). 2007. *The IPCC Fourth Assessment Report.* Cambridge: Cambridge University Press

19. Global CCS Institute. 2011. *The Global Status of CCS.* Details available at http://www.globalccsinstitute.com/publications/global-status-ccs-2011, p. 78

20. MacGill, I. F., R. J. Passey, and T. Daly. 2006. The limited role for carbon capture and storage (CCS) technologies in a sustainable Australian energy future. *International Journal of Environmental Studies* 63:752–763

21. Herzog, H., E. Drake, J. Tester, and R. Rosenthal. 1993. *A Research Needs Assessment for the Capture, Utilization, and Disposal of Carbon Dioxide from Fossil Fuel-Fired Power Plants.* U.S. Department of Energy, Washington, DC, DOE/ER-30194

22. National Energy Technology Laboratory. Details available at http://www.netl.doe.gov/

23. Warburton, Catherine, Andrew Gilder, Sibusiso Shabalala, and Melissa Basterfield. 2007. *Greenhouse gas mitigation mechanisms: South African policy & strategy and lessons from international jurisdiction: Companion Resource for BASIC Paper 12*, September

24. South African Centre for Carbon Capture and Storage. 2013. Impacts of CCS on South African national priorities other than climate change. *SLR Report*, SACCS Ref 6950-845-90-111-SA-SA I, October

25. Large Scale Integrated Projects in China. 2011. Global Status of Large-Scale Integrated CCS projects, Global CCS Institute, December Update

26. Gu, Yan. 2013. Carbon Capture and Storage Policy in China. Centre for Climate Change Law, *White Paper*, Columbia Law School

27. Climate Change Policy & Practice. 2012. ADB supports China's CCS Roadmap 2012. Details available at http://climate-l.iisd.org/news/adb-supports-chinas-ccs-roadmap/

28. Goel, Malti. 2004. Promoting technology development through inter-sectoral partnership. *Tech Monitor, PPP for Technology Development and Commercialization* July–Aug, pp. 32–38

29. India: Greenhouse Gas Inventory 2007. INCCA- Indian Network for Climate Change Assessment, 2010, Ministry of Environment and Forests, Government of India. Details available at http://www.moef.nic.in/downloads/public-information/Report_INCCA.pdf

30. IRADe and ICF International. 2010. *Analysis of GHG Emissions for Major Sectors in India: Opportunities and Strategies for Mitigation.* Washington, DC; Center of Clean Air Policy

31. Shackley, S. and P. Verma. 2008. Tackling CO_2 reduction in India through use of CO_2 capture and storage (CCS): Project and challenges. *Energy Policy* 36:3554–3561

32. Shahi, R. V. 2010. Carbon capture and storage technology: A possible long-term solution to climate change challenge. In *CO_2 Sequestration Technologies for Clean Energy*, S. Z. Qazim and Malti Goel (eds), pp.13–22. Delhi: Daya Publishing House

33. Goel, Malti. 2008. Carbon capture and storage, energy future and sustainable development: Indian perspective. In *Carbon Capture And Storage R&D Technologies for Sustainable Future*, Malti Goel, B. Kumar, S. Charan, and Nirmal (eds). Delhi: Narosa Publishing House

34. Ministry of Finance. 2010. Proactive steps in Budget 2010-11 for the environment. Details available at http://pib.nic.in/newsite/erelease.aspx?relid=58419

35. Strategic analysis of the global status of carbon capture and storage. 2009. Report 4: Existing Carbon Capture and Storage Research and Development Networks around the World. *Global CCS Institute*, Final Report

36. TERI 2013. *India CCS Scoping Study: Final Report.* Prepared for Global CCS Institute, January

37. Prathap, Gangan. 2010. How much a nation should spend on academic research? *Current Science* 98:1182–8

38. Pietzner, Katja Diana Schumann, and Sturle D. Tvedt. 2010. Scrutinizing the impact of CCS communication on the general and local public. *Report by Wuppertal Institute for Climate, Environment and Energy*, Germany, March

39. Brendan, Beck and J. Gale. 2009. Improving the global carbon capture and storage educational capacity. *Energy Procedia* 1:4727–33

40. Goel, Malti. 2009. Awareness and capacity building programme on carbon capture and storage. *Current Science* 98:606–607

41. Goel, Malti. 2009. Recent approaches in CO_2 fixation research in India and future perspective towards zero emission coal based power generation. *Current Science* 97:1625–1633

CHAPTER 4

TERI's Scoping Study on Carbon Capture and Storage in the Indian Context

Agneev Mukherjee, Arnab Bose, and Amit Kumar

The Energy and Resources Institute (TERI), India Habitat Centre, New Delhi-110003

E-mail: akumar@teri.res.in

INTRODUCTION

Carbon capture and storage (CCS) refers to the separation of carbon dioxide (CO_2) from industrial and energy-related sources, transport to a storage location, and long-term isolation from the atmosphere [1]. It is one among the portfolio of measures being considered for reducing greenhouse gas (GHG) emissions with a view to mitigating climate change. The importance of CCS is based predominantly on the fact that fossil fuels are expected to provide a large portion of the world's primary energy in the decades to come and, therefore, a means of reducing emissions from large-scale users of fossil energy will be necessary. CCS may be a viable option to this end. The International Energy Agency (IEA) has estimated that reducing GHG emissions by 50% by 2050 in the most cost-effective manner will require CCS to contribute about one-fifth of the necessary emissions reductions [2].

The three major aspects of CCS are CO_2 capture including its separation from the other gases produced in power plants, cement plants, steel mills, or other concentrated emission sources; transport after purification of the captured CO_2 by pipelines, ships, or other means to a site suitable for geological storage; and long-term storage via injection of the CO_2 into secure rock formations with monitoring required to guard against its leakage back into the atmosphere. Although various technologies for each stage have been independently available commercially for years, they are presently competing to be the low-cost solution by integrating them into CCS chain [2].

To understand the role that can be played by CCS in an Indian context, The Energy and Resources Institute (TERI) recently conducted a scoping study for CCS in India with support from the Global Carbon Capture and Storage Institute (GCCSI). The conclusion of the study arrived at through an examination of issues, opportunities and barriers to the deployment of CCS in the country, should help in drawing a road map for CCS implementation in India. The main findings of the study are presented here.

COUNTRY BACKGROUND

India's CO_2 Emissions

India's total GHG emissions in 2007 inclusive of land use, land-use change, and forestry (LULUCF) were 1727.71 million tonnes (Mt) of CO_2 equivalent and gross CO_2 emissions were 1497.03 Mt. The CO_2 generation per capita was 1.3 tonnes/capita, when not considering LULUCF [3]. Around 66% of India's gross CO_2 emissions came from the energy sector in 2007 with electricity generation alone accounting for almost 48% of the gross emissions. The industrial sector accounted for most of the remaining CO_2 emissions with 27% of the total emissions.

As per India's Integrated Energy Policy [4], India's CO_2 generation in 2031–32 is expected to be in the range of 3.9–5.5 billion tonnes (Bt) depending on India's economic growth, energy and carbon intensity of the economy, the share of renewables in India's energy mix, and other factors. This when combined with India's estimated population of 1468 million in that year means that India's per capita CO_2 emissions in 2031–32 are projected to be between 2.6–3.6 tonnes/capita. While the precise proportion of the emissions contributed by the various sectors will depend on the assumptions, the share of electricity generation is expected to continue to account for a majority of CO_2 emissions.

Current Climate Change Policies

India recognizes that it is highly vulnerable to climate change and hence ready to be a part of the solution. India announced in 2009 that it will reduce the emissions intensity of its gross domestic product (GDP) by 20%–25% over the 2005 levels by 2020 [5].

To address the climate change issue, the Indian Prime Minister's Council on Climate Change released the National Action Plan on Climate Change (NAPCC) in 2008. It outlines how despite not having any fixed, legally binding emission reduction targets as it is a non-Annex I country,

India still takes the issue of global warming seriously given that the government expenditure on climate change adaptation already exceeds 2.6% of GDP and that climate change is expected to have major impacts on water resources, agriculture, forests, and so on [6]. It explains how India's development will be on a sustainable trajectory stating that, "India is determined that its per capita GHG emissions will at no point exceed that of developed countries even as we pursue our development objectives."

Accordingly, eight national missions for managing climate change have been set up, which are the National Solar Mission; the National Mission for Enhanced Energy Efficiency; the National Mission on Sustainable Habitat; the National Water Mission; the National Mission for Sustaining the Himalayan Ecosystem; the National Mission for a 'Green India'; the National Mission for Sustainable Agriculture; and the National Mission on Strategic Knowledge for Climate Change. The Principal Scientific Advisor has announced the government's interest in adding a ninth 'Clean Coal Technologies' mission which would include CCS.

Major CO_2 Emitting Sectors

The major Indian CO_2 emitting sectors are described in brief below:

Power generation

India's installed electricity generation capacity has reached 210.9 GW [7] (in 2012). Coal accounts for 57.3% of the installed capacity with its share in the actual generation of electricity even higher owing to the low plant load factors that renewable power sources have as compared to thermal power plants.

Coal is expected to remain the mainstay of India's power sector in the near future too because most of the 100 GW of power capacity addition planned in the 12th Five-Year Plan period (2012–17) is based on coal-based power. However, future capacity addition is expected to be increasingly based on supercritical technology with 50% of the capacity in the 12th Five-Year Plan period targeted to be through supercritical units and all coal-based plants in the 13th Five-Year Plan period to be supercritical units [8].

Oil and gas production

India has recoverable crude oil reserves of 757.44 Mt [9], which is inadequate to meet India's growing energy needs with the result that the gap between domestic oil production and consumption has steadily been increasing. This shortfall is met by imports with 2.2 million barrels per day imported in 2010.

India's natural gas reserves were estimated to be 1241 billion cubic metres (BCM) in 2010 [9]. Consumption outstrips production with 2010 figures of 65 BCM for annual consumption and 51 BCM for annual production [10]. Presently, the power and fertilizer sectors together account for nearly three quarters of India's natural gas consumption and demand in the power sector is expected to grow in the future.

It should be mentioned that the oil and gas sector is significant not just as a source of CO_2 emissions, but possibly also as a sink via enhanced oil recovery (EOR).

Cement

India's strong economic growth in the recent past has coincided with an infrastructure boom in the country leading to the cement industry recording a compound annual growth rate (CAGR) of 8.1% over the last decade [11], which in absolute terms reflects the addition of 100 Mt capacity addition between 1999 and 2009 [12]. Today, the Indian cement industry is the second largest in the world with an installed capacity of 323 Mt.

The Indian cement industry is one of the most energy efficient in the world with the clinker plants having the lowest final energy use (3.1 GJ/t of clinker) and the specific electricity consumption also being the lowest in the world (~90 GJ/t for grinding) [13]. Alongside other improvements in technology and energy management, an important factor in this achievement is the fact that decades ago the industry started moving from the wet to the dry process for cement manufacture with the result that the proportion of the dry process has increased from 1% of cement production in 1960 to 97% in 2008 as against a decline in the share of the wet process from 94% to 2% in the same period [14].

One of the outcomes of the high energy efficiency of the Indian cement industry is the fact that average CO_2 emissions for the sector are among the lowest in the world at 0.68 Mt CO_2/Mt cement as compared to a global average of 0.84 Mt CO_2/ Mt cement [15]. The sheer volume of cement production, however, means that the industry emitted 129.92 Mt of CO_2 in 2007, a figure that may increase substantially in the future given India's aim of increasing cement manufacturing capacity to 479 Mt by 2017 [16].

Iron and steel

The Indian iron and steel industry has grown rapidly over the past two decades and today, it is the fourth largest producer of crude steel and the largest producer of sponge iron in the world. In 2009–10, India's production of pig iron was 5.88 Mt, while that of sponge iron was 24.33 Mt [17]. The total finished steel production in the same year was 60.62 Mt, which can be contrasted with a production of 14 Mt in 1991 [18]. This rapid growth rate is expected to be sustained in the near future with the Working Group on Steel for India's 12th Five-Year Plan projecting that India's crude steel capacity is likely to be 140 Mt in 2016–17 [17].

The average CO_2 emissions intensity of Indian steel plants at 2.4 t of CO_2/t finished steel is considerably higher than the global average of 1.8 t CO_2/t steel [19]. The fact that the emission intensity of India's iron and steel sector is much higher than the world standards means that CCS may be an attractive option for bringing this figure down to more acceptable levels.

Current CCS activity in India

Most Indian research and development (R&D) activities related to CCS occur under the Department of Science and Technology (DST) of the Indian Ministry of Science and Technology [20]. The National Programme on Carbon Sequestration (NPCS) Research was set up by DST in 2007 with a view to compete with other countries in this area with respect to both pure/applied research and industrial applications. Four thrust areas of research were identified under this programme namely, CO_2 sequestration through microalgae bio-fixation techniques; carbon capture process development; policy development studies; and network terrestrial agro-forestry sequestration modelling [21]. Information on various projects can be found in DST annual reports [22–23]. In addition, state-owned entities such as Oil and Natural Gas Corporation (ONGC), National Aluminium Company (NALCO), and National Thermal Power Corporation (NTPC) have also been investigating different aspects related to CCS as have been research organizations like Indian Institute of Technology (IIT), Bombay and Indian Institute of Petroleum (IIP), Dehradun.

ECONOMIC ANALYSIS

Given that a majority of India's emissions come from the power sector and that the development of gigawatt scale power plants in recent years means that the large-scale concentrated emission sources that are most suitable for CCS deployment are predominantly in this sector, it is a gigawatt scale power plant that has been considered for economic analysis in the study. It was assumed that the plant is built capture ready as per the following definition [24]:

"A CO_2 capture ready power plant is a plant, which is initially not fitted with CO_2 abatement technology, but which subsequently can be fitted with a technology to capture the gas, when regulatory or economic drivers are in place to drive this."

Three different cases with a different set of assumptions were analysed and it was concluded that CCS deployment will lead to an increase in the levelized cost of electricity (LCOE) of about 38%–47% over the base plant cost. The two principal reasons for the increase are the high capital cost of the capture equipment, which amounts to more than 60% of the base plant capital cost, and the capture energy penalty, which leads to an increase in the fuel required to generate the same net amount of power.

POLICY AND LEGISLATION REVIEW

Legislation that may govern CCS activities is limited mostly to the following sectors:

Oil and Gas

- Indian Petroleum Act, 1934: Rules for production and transportation of petroleum products. It can be applied for transportation of compressed CO_2.
- The Oilfields (Regulation and Development) Act, 1948 (53 of 1948): Royalties in respect of mineral oils. It can be applied for EOR.
- The Petroleum Mineral Pipelines (Acquisition of Right of User in Land) Act, 1962: Provides for the acquisition of user in land for laying pipelines for the transport of petroleum and minerals and for matters connected therewith. This law may be applied for transportation of compressed CO_2 to storage sites.
- The Oil Industry (Development) Act, 1974: An act to provide for the establishment of a board for the development of oil industry and for that purpose to levy a duty of excise on crude oil and natural gas and for matters connected therewith. It can be modified for levying a duty of excise on crude oil and natural gas produced during EOR.

- Petroleum and Natural Gas Rules, 1959: An act to provide petroleum exploration license and mining leases. This law will be for development of sites for EOR and enhanced gas recovery (EGR).
- Directorate General of Hydrocarbon (DGH): Under Ministry of Petroleum and Natural Gas, Government of India is looking after development of coalbed methane (CBM) production. This may become relevant to enhanced CBM.

Groundwater

Water (Prevention and Control of Pollution) Act 1974 enacted by Ministry of Environment and Forest, Government of India provides for the prevention and control of water pollution and for maintaining or restoring of wholesomeness of water in the country [25]. This act levies and collects cess on water consumed by persons operating and carrying on certain types of industrial activities. CCS has environmental impacts in terms of chances of groundwater contamination and this act could be suitably modified to include contamination of groundwater in case there is any leakage of stored CO_2.

Environmental Impact Assessment

Amending the Environmental Protection Act, 1986 is likely to be the most-effective way to facilitate demonstration projects and may be done on a project-specific basis before broader amendments can be established. Since CO_2 may need to be transported across states and be stored in a region different to the point of collection, regional coordination groups will need to be established to address issues related to CO_2 transport and storage. Retrofitting of CO_2 capture capability to existing power plants may be done under the Environment Impact Assessment Notification S.O.60 (E) under the provisions of the Environment (Protection) Act 1986.

Barriers to CCS Implementation in India

In the following, we summarize the principal barriers for CCS deployment in India that have been raised by some Indian stakeholders:

- Worldwide, CCS is still in the demonstration phase. Once a degree of confidence has been gained in the technology via large-scale deployment internationally, then it can be considered seriously for India.

- One major barrier to CCS deployment in India is the lack of accurate geological storage site data, since, before capture technology can be installed in power plants or other sources, the location, capacity, permeability, and other characteristics of the sinks must be known.
- The issue of CCS drastically increasing the cost of electricity, while reducing net power output is often cited as being one of the biggest barriers to acceptability of CCS in India. CCS deployment is held to run counter to India's ambitious goals for electrification, especially given the present electricity deficit and energy situation in the country.
- EOR is considered as one of the most attractive options for CO_2 storage, since the cost of storing CO_2 is partially offset by the revenues accrued by the hard-to-extract oil that can be recovered from depleted oilfields. In the Indian scenario, however, it has been stated by stakeholders in the petroleum sector that there are few oilfields, which are sufficiently depleted to be relevant for EOR at present; further, since EOR is dependent on the miscibility characteristics of the oil with the extracting fluid, it may not be suitable in all the cases.
- Clarity is needed on how CCS implementation via retrofit of capture equipment to existing plants will change the terms of reference (ToR) of the power plant. In particular, the fresh environment clearances required, if any, need to be spelt out and standardized.
- Access to funding from financing agencies such as the World Bank, Asian Development Bank and so on, might require further governance requirements in addition to the existing requirements, for example, around monitoring, measure, and verification. These may be dependent on CCS-specific clearances being available from the Ministry of Power and/or other government bodies, in addition to the existing clearances required.
- Deployment of CCS on a large scale requires specialized manpower and suitable infrastructure, which may not be available in India at present.
- Monitoring the stored CO_2 to assure against leakage is essential if the central purpose of CCS implementation is to be fulfilled. Ensuring that rigorous monitoring is needed over long-time duration. The techniques developed internationally in this area need to be introduced to Indian stakeholders.

- Legal issues related to land acquisition, groundwater contamination, CO_2 leakage and so on, need to be addressed before any large-scale transport and storage of CO_2 can be permitted.

CAPACITY DEVELOPMENT NEEDS

Broadly, the following capacity development needs related to CCS require to be addressed, so as to create an enabling environment for CCS deployment in India:

- Knowledge building and capacity development of policy makers and regulators
- Capacity development for storage site assessment, development, operation, and monitoring and verification
- Technology sharing and transfer
- Capacity development of financial institutions
- Public engagement
- Knowledge sharing among different CCS groups

CONCLUSION AND RECOMMENDATIONS

Presently, India's top development priority is to provide electricity to all at affordable prices. Given the relative abundance of India's coal reserves, it is natural that coal is the predominant constituent of India's power sector, a status that is likely to continue for decades to come. Despite this, India is exploring all possible means of reducing GHG emissions, including increasing the use of supercritical technology for power generation. India also supports global efforts at R&D into CCS technologies. However, there are stakeholder concerns pertaining to the capital and operating costs; the energy penalties; and safety and integrity of potential storage; and the social acceptance of CCS. The high cost of electricity and reduced net electricity generation with CCS challenges the country's goal of 'electricity to all and at affordable prices'. But given its wide applicability, the role of CCS is not limited to power generation alone, but extends to other industrial sectors as well including utilization of captured CO_2 for EOR, manufacture of cement substitutes, algal biofuel, fertilizer manufacture, and mineralization. Indeed, the role of CCS as a potential climate change mitigation option for India needs to be explored further, so that issues raised in this study could be suitably addressed.

Some recommendations for addressing these issues are made as follows:

- For CCS to be considered as a viable mitigation option in India, a major challenge is the lack of reliable storage data. The relevant Indian institutions and organizations need to work together for the preparation of an accurate geological storage map for India utilizing recognized assessment criteria and incorporating the different storage options such as saline aquifers, basalt rocks, and depleted oilfields (on-shore and off-shore). Finally, a matching of sources and sinks and cost optimization in this regard will need to be carried out.

- The overall cost of capturing and storing CO_2 has not been quantified accurately for an Indian scenario so far and, hence, a life cycle analysis (LCA) of the entire system needs to be conducted.

- The whole issue of financial risks and legal liabilities in the case of CO_2 leakage from the storage site needs to be addressed appropriately.

- R&D in the areas of improved capture systems, more efficient retrofit, and plant integration should be undertaken for increasing CCS acceptability.

- To facilitate better interaction with the global CCS community, it may be worthwhile to devise a mechanism for knowledge and experience sharing among different actors namely, the technology developers/suppliers, global practitioners of CCS and potential users, research community, as well as decision-makers from India on different aspects of CCS on a regular basis. Such interactions would also be helpful from the policy planning perspective.

- Sustained efforts are required towards capacity development of different stakeholders including sensitization of the policy makers and the regulators about the latest developments in this field.

- Public acceptance of CCS being central to its successful deployment, workshops and seminars disseminating information about this technology may be conducted to increase mass awareness.

Summary

In this chapter TERI presents results of a scoping study carried out on the present status and potential applications of CCS in India. Main results of the scoping study are presented. The study reveals the main areas in which CCS deployment could be considered in India and highlights the main barriers that will need to be overcome along with the recommendations made in this regard.

REFERENCES

1. Intergovernmental Panel on Climate Change (IPCC). 2005. *Special Report on Carbon Dioxide Capture and Storage.* Cambridge: Cambridge University Press

2. International Energy Agency (IEA). 2009. *Technology Roadmap: Carbon Capture and Storage.* Paris: International Energy Agency

3. Ministry of Environment & Forests, Government of India. 2010. *India: Greenhouse Gas Emissions-2007.* New Delhi: Ministry of Environment and Forests

4. Planning Commission, Government of India. 2006. *Integrated Energy Policy: Report of the Expert Committee.* New Delhi: Planning Commission

5. Planning Commission, Government of India. 2011. *Low Carbon Strategies for Inclusive Growth: An Interim Report.* New Delhi: Planning Commission

6. Prime Minister's Council on Climate Change, Government of India. 2008. *National Action Plan on Climate Change.* Details available at http://www.moef.nic.in/modules/about-the-ministry/CCD/NAP_E.pdf

7. Central Electricity Authority, Government of India. 2012. *Monthly Report on Installed Capacity.* Details available at http://www.cea.nic.in/reports/monthly/inst_capacity/nov12.pdf

8. Press Information Bureau, Government of India. 2010. Power Minister's Address at EEC. Details available at http://pib.nic.in/newsite/erelease.aspx?relid=66577

9. Ministry of Petroleum & Natural Gas, Government of India. 2011. *Basic Statistics on Indian Petroleum & Natural Gas 2010–11.* New Delhi: Ministry of Petroleum & Natural Gas

10. Energy Information Administration, Department of Energy, USA. 2011. *Country Analysis Briefs: India.* Details available at http://www.eia.gov/countries/country-data.cfm?fips=in

11. Confederation of Indian Industry. 2012. *Cement Industry in India: Trade Perspectives.* Details available at <http://newsletters.cii.in/newsletters/mailer/trade_talk/pdf/Cement%20Industry%20in%20India%20Trade%20Perspectives.pdf>

12. Department Related Parliamentary Standing Committee on Commerce, Parliament of India. 2011. *Ninety Fifth Report on Performance of Cement Industry.* New Delhi: Rajya Sabha Secretariat

13. United Nations Industrial Development Organisation. 2010. *Global Industrial Energy Efficiency Benchmarking.* Vienna: United Nations Industrial Development Organisation

14. Saxena, A. 2010. Best practices and technologies for energy efficiency in Indian cement industry. *National Council for Cement and Building Materials.* Details available at http://www.iea.org/work/2010/india_bee/saxena.pdf

15. Bhushan, Chandra. 2009. *Challenge of the New Balance.* New Delhi: Centre for Science and Environment

16. Press Trust of India. 2012. India's cement-making capacity pegged at 479 Mt by 2017. *The Economic Times,* 8 January

17. Ministry of Steel, Government of India. 2012. *An Overview of Steel Sector.* Details available at <http://steel.nic.in/overview.htm>

18. Corporate Catalyst India. 2012. *Iron and Steel Industry in India.* Details available at http://www.cci.in/pdf/surveys_reports/iron-steel-industry.pdf

19. GIZ. 2011. *Steel Industry in India: Potential and Technologies for Reduction of CO_2 Emissions.* Details available at http://www.hrdp-net.in/live/hrdpmp/hrdpmaster/hrdp-asem/content/e18092/e21298/e25159/e40403/eventReport40420/Presentation-GIZsteelindustryworkshopreport_Corr1_EditedNP-100112-2.pdf

20. Goel, Malti. 2009. Recent approaches in CO_2 fixation research in India and future perspective towards zero emission coal based power generation. *Current Science* 97:1625–1633

21. Department of Science and Technology, Government of India. 2008. *Joint Technology Projects under STAC/IS-STAC.* Details available at http://www.dst.gov.in/about_us/ar07-08/tech-dru-prg.htm#stac

22. Department of Science and Technology, Ministry of Science and Technology, Government of India. 2010. *Annual Report 2009–10.* Details available at <http://www.dst.gov.in/about_us/ar09-10/annual_report_2009-10.pdf>

23. Department of Science and Technology, Ministry of Science and Technology, Government of India. 2012. *Annual Report 2011–12.* Details available at http://www.dst.gov.in/about_us/ar11-12/PDF/DST_Annual_Report_2011-12.pdf

24. MacDonald, Mott. 2008. *CO_2 Capture-Ready UMPPs, India.* Details available at https://ukccsrc.ac.uk/system/files/publications/ccs-reports/DECC_CCS_52.pdf

25. Ministry of Environment and Forests, Government of India. 2012. *Water Pollution.* Details available at <http://moef.nic.in/modules/rules-and-regulations/water-pollution/>

Section II
CARBON CAPTURE AND STORAGE

CHAPTER 5

Capturing CO_2 by Physical and Chemical Means

A. K. Ghoshal[1], P. Saha[2], B. P. Mandal[2], S. Gumma[3], and R. Uppaluri[2]

[1]Professor, Department of Chemical Engineering and Head, Centre for Energy
[2]Professor, Department of Chemical Engineering
[3]Associate Professor, Department of Chemical Engineering,
Indian Institute of Technology Guwahati,
Guwahati-781039, Assam, India
E-mail: aloke@iitg.ernet.in

INTRODUCTION

The sectors such as energy, agriculture, automobile, industry, and wastes are mainly responsible for the emissions of greenhouse gases (GHGs). In the energy sector, fossil fuel combustion plays a major role in GHG emissions by adding carbon dioxide (CO_2) and contributing to global warming. The total global energy use by 2025 is expected to rise by 75% of the energy use pattern of 1996 and consequently CO_2 emissions would increase. The primary objective of this chapter is to focus on the CO_2 capture technologies. The present research and development (R&D) activities are directed to refinement of current capture technologies on the one side and development of novel capture technologies that can deliver significant benefits on the other side. Figure 1 indicates possible capture and fixation routes for CO_2 and other GHGs. The CO_2 capture is dealt into two categories viz., post-combustion capture and pre-combustion capture. In the former case, CO_2 is captured from an exhaust gas (typically at atmospheric pressure) by passing through a separation unit, which may work on absorption or adsorption. In an alternate process (oxy-fuel process: mostly in R&D stage), the fuel is burnt in an oxygen-rich environment so that mostly CO_2 is emitted and the need for CO_2 separation is eliminated. In pre-combustion capture, the fuel is not burnt directly, but is converted at suitable temperature

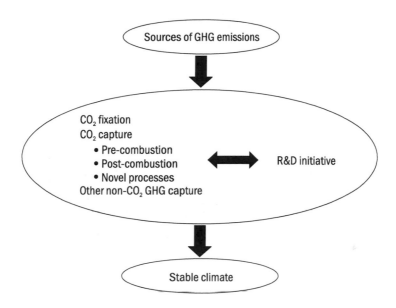

Figure 1   *Capture and fixation technologies for CO_2 and other GHGs*

and pressure into synthesis gas (syn-gas) [mixture of carbon monoxide (CO), CO_2, and hydrogen (H_2)]. Thereafter, CO is further converted into CO_2 and H_2, and then CO_2 is captured to get H_2 (the major constituent) as fuel. Membrane Reforming Sorption Enhanced Water-Gas-Shift (MRSEWGS) with Integrated Gasification Combined Cycle (IGCC) is a typical example of pre-combustion capture technology. Fundamentals of these technologies, the state-of-the-art research, on-going and completed research/commercial projects are discussed here.

POST-COMBUSTION CARBON CAPTURE

Post-combustion carbon capture mainly involves capture of CO_2 from flue gas streams. Chemical and/or physical absorption, physical adsorption, and membrane separation are typically the available significant CO_2 capture technologies [1].

CO_2 Capture by Absorption

Bottoms [2] in 1930 demonstrated the process of absorption to remove CO_2 using amines, which is the most widely practiced CO_2 capture process at present. Absorption can be both physical absorption and chemical absorption. In physical absorption, CO_2 gets soluble in the absorbing liquid (absorbent, water, and specialty physical solvents) and does not react with the absorbent, thus, the equilibrium concentration of the absorbate, CO_2 in the liquid phase is strongly dependent on its partial pressure in the gas phase. Chilled methanol (Rectisol process), N-methyl-2-pyrolidone (Purisol process), dimethylether of polyethylene glycol (Selexol process), propylene carbonate (Flour process), and so on are typically used physical absorbents. Though, physical solvents exhibit good equilibrium loading capacity with easy regeneration and are useful at relatively high partial pressure of CO_2 (above 1.3 MPa), they are expensive. Therefore, chemical solvents are preferred when processing high throughput streams (typically, greater than 6.6 m³/s) due to their higher absorption rate, even though the regeneration cost is higher. Chemical absorption involves a chemical reaction between CO_2 and a component of the liquid phase [hot potassium carbonate solution, ammonia (NH_3), and amine in aqueous and non-aqueous solvents and so on] to form a loosely bound reaction product [3]. The most widely used chemical solvent processes are aqueous alkanolamine and promoted hot potassium carbonate processes.

At present, less than 100 physical solvent processes are in place compared to 2000 installations using chemical solvents [4]. A schematic diagram of the typical process for CO_2 capture by chemical absorption is shown in Figure 2. The major absorption processes currently in practice are discussed in the ensuing paragraphs.

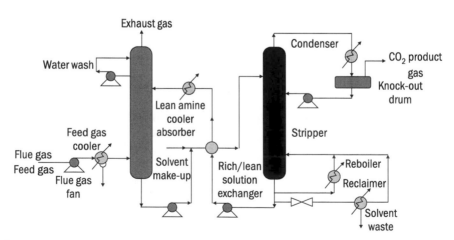

Figure 2 *Typical schematic diagram for CO₂ capture by chemical absorption*

Aqueous alkanolamine solutions for CO_2 capture

Most of the acid gas treating processes, in operation today, use alkanolamine solvents because of their versatility, ability to remove to very low-levels and lower cost of aqueous solutions [4]. In addition to monoethanolamine (MEA: primary amine), diethanolamine (DEA: secondary amine), methyldiethanolamine (MDEA: tertiary amine) and so on, several proprietary formulations of alkanolamine solutions containing corrosion inhibitors, foam depressants, and activators are being used under various trade names such as UCARSOL, Amine Guard (Union Carbide Corporation), GAS/SPEC IT-1 Solvents (Dow Chemical Company) and Activated MDEA (BASF Aktiengesellschaft), KS-type solvents, and so on.

Primary and secondary amines present in water react rapidly with CO_2 to form carbamate ions enhancing the CO_2 absorption capacity and rate, but the cost of regenerating the amines is high. They have the disadvantage of limited loading as two moles of amine are required to react with one mole of CO_2. Tertiary amines do not react directly with CO_2, but promote the hydrolysis of CO_2 to form bicarbonate and protonated amine, which is much slower than the direct reactions of primary and secondary amines with CO_2. The kinetic selectivity of tertiary amines towards CO_2 is poor, but the regeneration cost is lower. In addition, tertiary amines allow for very high equilibrium loadings as the reaction stoichiometry with CO_2 is 1:1. Thus, primary and secondary alkanolamines show kinetic selectivity and tertiary alkanolamines show thermodynamic selectivity towards CO_2.

The other category, the sterically hindered amines, can be primary (such as 2-amino-2-methyl-1-propanol, AMP) or secondary (such as diisopropanolamine, DIPA), exhibit highly reversible kinetics with CO_2, and thus require less energy for regeneration. In addition, besides saving energy and capital in gas treating processes significantly, the hindered amines show better stability than conventional amines, because of low amine degradation. Mitsubishi Heavy Industries' (MHI) KS-1, KS-2, and KS-3 solvents have shown promise for CO_2 capture due to high CO_2 loading, low solvent degradation, low corrosion, and low steam consumption in the stripper. PETRONAS plant in Malaysia is based on KS-1 solvent. In recent years, blended amine solvents (typically mixture of primary or secondary amine with a tertiary amine or sterically hindered amine) are employed since they combine the higher CO_2 reaction rates of the primary or secondary amine with the higher CO_2 loading capacity of the tertiary amine or sterically hindered amine. Thus, lower solvent circulation rate is necessary compared to a single amine solvent, which is one of the most important factors in determining the economics of a CO_2 capture process using chemical solvent [5].

Solvent circulation rate plays an important role in deciding the size of absorber, pump, pipeline, and heat exchanger and, hence, the capital cost of the capture plant. The reboiler heat duty is also correlated with liquid circulation rate. Lower circulation rate requires less regeneration energy, which accounts for about 70% of the total operating cost of a gas treating process [4].

Carbonate process

The hot potassium carbonate process originally developed during 1950–70 requires relatively high partial pressures of CO_2 (above 215 k Pascal). There is a large similarity between the process flow schemes for the hot potassium carbonate and the amine processes [4]. In this process, gas containing CO_2 is contacted with an aqueous solution of potassium carbonate, where CO_2 reacts and forms potassium bicarbonate and CO_2 free gas is separated. The liquid is regenerated upon heating. Increased temperature enhances the rate of reaction. The process also readily removes 90% of hydrogen sulphide from the feed gas as the former reacts with potassium carbonate solution. Proprietary activators and inhibitors are used for enhancing CO_2 mass transfer and for reducing corrosion. Typical applications of the process includes CO_2 removal in NH_3 plants, direct iron ore reduction plants, natural gas treatment, and recycle gas purification in ethylene oxide facilities. This technology has been used in Indian Farmers Fertilizer Cooperative Ltd's commercial fertilizer plant in Phulpur, India. The major disadvantages include high CO_2 partial pressure requirement, large heat requirements for regeneration, as well as foaming and corrosion.

Chilled NH₃ process

In chilled NH_3 process, the flue gas is cooled to very low temperature (0°C–10°C), which reduces the flue gas volume and increases CO_2 concentration as well. At about the same temperature, the absorber is operated to have high CO_2 capture efficiency and low NH_3 emission. Regeneration is carried out at about 120°C and 20 bar to generate high pressure CO_2 with low concentrations of moisture and NH_3. The advantages of NH_3-based process are: it uses less expensive solvents to remove CO_2; low partial pressure CO_2 can be treated; energy requirement is also smaller compared to conventional MEA-based processes [4]. However, chilled NH_3 process has the disadvantages of higher corrosivity of the loaded solution and complex flow scheme compared to the alkanolamine and hot potassium carbonate processes.

CO_2 capture from power plants

Monoethanolamine, because of its ability to absorb CO_2 under low partial pressure, has been used for the separation of CO_2 from power plant flue gases. Various MEA-based plants, for example Tokyo Electric Power Company, Kansai Electric Power Company, Kerr-McGee Chemical Company, have been installed to separate CO_2 from flue gases. Table 1 indicates that MEA costs better than cryogenic fractionation and membrane separation techniques for various CO_2 separation schemes compared in terms of kg of coal required per kWh of power generation for the base case (without CO_2 separation) with CO_2 separation using different schemes [6]. The scenario is likely to change for Indian conditions due to difference in the quality of coal. However, the significant amount of energy requirement for the MEA process is primarily for its regeneration, which demands for further research to find energy efficient process that can replace the MEA process [6].

In the absorption processes, the solvent activity must be balanced between absorption and desorption rates so that the energy penalty is as little as possible. High oxygen content of flue gas can be corrosive particularly in high temperature stripping section and can also cause excessive amine loss due to degradation. Suitable sterically hindered amines offer resistance to such degradation. It is preferable to keep levels of oxides of sulphur (SO_X) below 0.001% to avoid formation of stable salts with the amines used for absorption. Thus, an SO_X scrubber is generally a cost-effective option. Fly ash and oxides of nitrogen (NO_X)

Table 1 : Energy requirements for various CO_2 removal schemes [6]

Scheme	Coal requirements (kg/kWh)
No CO_2 removal	0.35
Amine scrubbing	0.66–1.68
Cryogenic fractionation	0.78–7.04
Membrane separation	0.72–1.41

compounds also create similar problems as SO_X. However, Indian coal has the advantage of low sulphur, and nitrogen contents though it suffers heavily from high ash content. In the case of Indian power plants, since SO_X emissions are within current acceptable limits, additional capital cost in the form of a Flue Gas Desulphurization (FGD) unit is not necessary. But this becomes necessary, when amine-based CO_2 capture unit is to be installed.

CO_2 CAPTURE BY ADSORPTION

Adsorption, a fluid-solid interaction phenomenon, has been successfully employed for various catalytic fluid-solid reactions and separation processes in chemical and petrochemical industries, biotechnology and environmental control engineering. For gas separation, adsorbate molecules by means of physical adsorption are held on the surface and in the pores of adsorbent by weak van der Waals' force of attraction. The process is rapid, reversible, and usually has low heat of adsorption [7]. Adsorption rates depend on temperature, partial pressures/concentration of adsorbate in bulk gas, surface forces (interaction energy between adsorbent and adsorbate), and pore size or available surface area of the adsorbent [1]. Pressure (or Vacuum) Swing Adsorption (PSA/VSA) and Temperature Swing Adsorption (TSA) are the main methods. Figure 3 is a typical schematic diagram showing adsorptive separation of CO_2 from a flue gas mixture. Separation is achieved by adsorption, because of steric

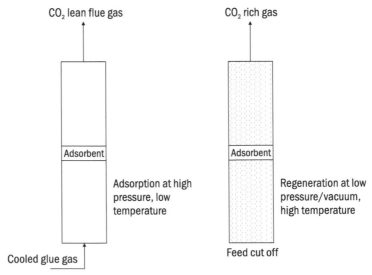

Figure 3 *Typical schematic diagram for CO₂ capture by adsorption*

effect (components differing in their molecular diameters and structures are separated); equilibrium effect (components utilizing the differences in adsorption capacities of the adsorbents for the adsorbates); and kinetic effect (components utilizing the differences in diffusive properties of the adsorbates in the pores of the adsorbents). Adsorbent provides the surface required for selective adsorption of the preferentially adsorbed species due to any or a combination of above effects. A large specific surface area is desirable. Since the capacity determines the size and, therefore, the cost of the adsorbent beds, high capacity is desirable alongside high selectivity, which is the primary requirement [8].

Most common physical adsorbents are microporous materials with high surface areas and specific sites of adsorption such as zeolites and ordered mesoporous silicas known as MCM-41, MCM-48, SBA-15, and SBA-16 [9]. Functionalizing these mesoporous materials with groups that promote CO_2 adsorption can enhance their capacity. The ordered mesoporous silicas have uniform and large pores, tunable pore sizes, high surface areas and large number of highly dispersed active sites (the hydroxyl groups) on the pore walls and surface. Most of the silanols/hydroxyl groups get removed during the calcinations step (KA KOH, a special type of scrubber is sometimes added to FGD scrubbers to reduce SO_X below acceptable limits, because the latter typically removes 90% of the SO_X in the flue gas). The residual hydroxyl groups are unable to induce strong enough interactions and, therefore, surface modification by adding functional groups (for example, amine tethering) for adsorption of CO_2 can increase the gas-adsorbent interactions. Knofel *et al.* [9] in the study with grafted amines onto surface SBA-16 concluded that such materials can be of potential interest to recover trace amounts of CO_2. In such functionalized ordered mesoporous adsorbents, performance is directly related to the amine content or the surface density of an amine.

Particle sizes of the adsorbents are also equally important, because of the adsorption dynamics. SBA-16 is an excellent mesoporous silica support due to its hydrothermally stable cubic cage structure with multi-directional and large pore systems that allow accessibility for functionalization and adsorption [10]. The maximum amount of CO_2 adsorbed on N-(2-aminoethyl)-3-aminopropyltrimethoxy silane (AEAPS) functionalized SBA-16 at 60°C was 0.727 mmol/g, where interaction between the amine group and CO_2 is chemisorption. Lithium silicate nanoparticles in microemulsion media have shown good performance for CO_2 adsorption at higher temperatures [11].

Activated carbons (ACs), activated clay, silica gel, and activated alumina are among other commercial adsorbents used for separating gas and vapour mixtures. They do not possess ordered crystalline structure and the pores are non-uniform. Table 2 represents typical characterization results of ACs [7, 12]. The adsorptive properties of micro-crystalline ACs are due to their microporous structure and high surface area [13–14]. Modification of surface chemistry of the ACs in addition to structural properties such as surface area and pore-size distribution can play an important role on adsorption capacity [15]. Modification can be done by creating acidic or basic groups.

In the case of CO_2, metal oxides, zeolites, and hydrotalcite compounds are also being tried for CO_2 separation.* In recent times, adsorption and separation of CO_2 has been widely studied on a variety of adsorbents [9–10, 16–19]. However, most of them do not have sufficiently high CO_2 adsorption capacity at high temperatures. The adsorption process of CO_2 capture from the flue gas of a power plant is described below.

CO₂ Capture from Flue Gas by Adsorption

Flue gases typically contain nitrogen, CO_2, hydrogen oxide, NO_x, SO_x, CO, oxygen, and particulate matter. CO_2 can be adsorbed in naturally occurring substances such as coal (a method of sequestering CO_2 in coal seams that cannot be mined) or synthetic adsorbents such as AC, zeolites, and so on. Application of adsorption for CO_2 capture from a variety of flue gas streams, however, suffers from the drawbacks so as to make it a stand-alone process. It does not handle large concentrations of CO_2, large volume of throughput, not selective enough for CO_2 separation and finally, adsorption is a slow process. Further research on more selective adsorbents with higher capacities and suitable operating conditions may make the adsorption a viable method in future.

Table 2 : Typical properties of AC [7]

Property	Micropore	Mesopore	Macropore
Diameter, Å	<20	20–500	>500
Pore volume (cm³/g)	0.15–0.5	0.02–0.1	0.2–0.5
Surface area (m²/g)	100–1000	10–100	0.5–2
Particle density: 0.6–0.9 g/cm³; porosity: 0.4–0.6			

---

* *Please see Chapter 6*

The commercial adsorptive separation processes like separation of landfill gas [CO_2/methane (CH_4) in PSA separation plant] by Air Products and Chemicals Inc. (APCI); Selexsorb adsorbent for the Natural Gas Combined Cycle (NGCC) by Alcoa and TSA technology by Air Liquide have been proposed. All these options suffer from the drawback of high value for cost of power. Among various capture technologies, PSA/VSA is becoming a promising option for separating CO_2 from flue gas due to its low operating and capital costs [20]. Further developments taking place are two-stage PSA [21] and dual reflux PSA process [22]. While comparing CO_2 separation from a low CO_2 concentration flue gas containing 17% CO_2 using AC and CMS (carbon molecular sieve), Chue *et al.* [23] pointed out that the equilibrium selectivity for CO_2 in AC dominates the PSA separation. They have also suggested that zeolite 13X gives more favourable results than AC. Tokyo Electric Power Company and Mitsubishi Heavy Industries Ltd employed Pressure and Temperature Swing Adsorption (PTSA) to separate and remove CO_2 from flue gases produced in coal-fired power plants with a target of 99% concentration of recovered CO_2 and a recovery ratio of 90% [24].

OXY-FUEL TECHNOLOGY

In oxy-fuel combustion, fuel is fired with an oxygen-enriched gas, which is produced (with 95% oxygen) by removing nitrogen from air, which is carried out with an Air Separation Unit (ASU). Table 3 indicates typical gas composition and volumes (omitting non-condensables) for bituminous coal ($CH_{1.1}$, $O_{0.2}$, $N_{0.017}$, and $S_{0.015}$) fired with air and oxygen. Advantages of oxy-fuel combustion are: production of a CO_2-rich flue gas ready for sequestration; reduction of mass and volume of the flue gas by approximately 75%; minimization of heat loss and large reduction of

Table 3 : Gas composition for bituminous coal ($CH_{1.1}$, $O_{0.2}$, $N_{0.017}$, and $S_{0.015}$) fired with air and oxygen [25]

Gas	Air firing	Oxy-firing
CO_2	17%	64% by volume
H_2O	8.9%	34%
NO_X	2,770f ppm	10,700f ppm
SO_X	2,470 ppm	9,400 ppm
Moles	1	0.26
f is fractional conversion of coal nitrogen to NO_X		

nitrogen dioxide production due to the presence of less nitrogen in the input gas. However, in this case, high flame temperature during firing with pure oxygen results can damage the process equipment. To prevent this, recycled flue gas (RFG) is mixed with oxygen-rich gas to dilute and, thereby, reduce the temperature. Temperature control is achieved by suitably adjusting recycle ratio.

Other merits of oxy-fuel combustion are that mass and volume of flue gas are reduced by approximately 75%, size of flue gas treatment equipment can be reduced by 75%, heat loss is less, flue gas containing primarily CO_2 is ready for sequestration, and NO_X production is greatly reduced.

Figure 4 gives a typical flowsheet of oxy-fuel combustion. The recirculation ratio, R depends upon furnace size, oxygen purity, and temperature of RFG. R should be around 3 if the heat flux for air combustion is to be matched. Though, it has been reported that NO_X and sulphur dioxide (SO_2) emissions reduce during oxy-firing conditions, the mechanism is yet to be fully understood. Following the work of Liu and Okazaki [26], the exit concentration of NO can be estimated from Equation 1 [25]:

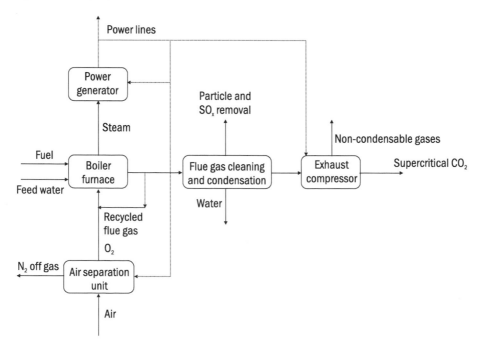

Figure 4 *Typical flowsheet of oxy-fuel combustion*

Table 4 : Percentage reduction of NO$_X$ with recycle ratio of RFG [25]

Recycle ratio of RFG	% reduction of NO$_X$
0	No reduction
2.25	69.5
2.75	78.5
3.25	82.5

$$(No)_{exit} = \alpha F_N / R_\eta + 1 \qquad (1)$$

where α is fraction of fuel nitrogen converted to nitric oxide (NO), F_N is fuel nitrogen, R is recycle ratio, and η is fraction of NO destroyed by re-burning. Table 4 represents typical percentage reduction of NO$_X$ against R.

The major applications of oxy-fuel technology in the last decade have been in power generation industries particularly for CO_2 reduction for IGCC and oxy-fuel boiler.

CHEMICAL LOOPING COMBUSTION

In a chemical looping combustion (CLC) technique, two fluidized beds (Figure 5) are usually employed. The metal reacts with air to form a metal oxide in the first bed, oxidizer. The metal oxide is then sent to the reducer (the second reactor), where fuel is injected. Fuel is oxidized (combusted) and the metal oxide is reduced to metal, which is again sent

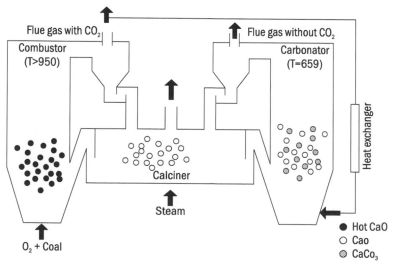

Figure 5 *Double calcium loop to capture CO_2 [31]*

back to the oxidizer for reuse, thus completing the loop. CO$_2$ produced by the exothermic reaction is first separated from the produced water, and then sequestered [27–30].

Though, the CLC technique was originally proposed to increase the combustion efficiency through appropriate oxygen carriers, this technique presently is being investigated in the laboratory stage as CO$_2$ capture method. Flue gas is usually free from NO$_X$ as there is no nitrogen in the reducer. National Energy Technology Laboratory (NETL), USA; Institute of Carbon Chemistry, Spain; Chalmers University of Technology in Sweden; Technical University of Vienna; BP in the UK and Alstom, France are the few groups and companies working on various projects with this technology. The major challenges in this technology include identification of suitable oxygen carrier (metal) to withstand several looping cycles without loss in performance and effective separation of metal and ash (if fuel is a solid like coal).

PRE-COMBUSTION CARBON CAPTURE METHODS

Pre-combustion capture technologies are: Hydrogen Membrane Reforming (HMR), Sorber Enhanced Water-Gas-Shift (SEWGS) Reaction, and IGCC. These are discussed as follows:

Hydrogen Membrane Reforming

Hydrogen membrane reforming technology also referred to as membrane water-gas-shift reaction (MWGSR) is a combination of steam reforming (SR) and water-gas-shift (WGS) reaction modelled into a single unit. The reactions are given by Equations (2), (3), and (4).

Steam reforming reaction: $CH_4 + H_2O \rightleftharpoons 3H_2 + CO$ (ΔH = 206 kJ/mol) (2)

Water-gas-shift reaction: $CO + H_2O \ H_2 + CO_2$ (ΔH = −41 kJ/mol) (3)

Overall reaction: $CH_4 + 2 H_2O \rightleftharpoons 4H_2 + CO_2$ (4)

Figure 6 is a typical membrane reactor, where natural gas and water are fed, both SR and WGS reactions occur in parallel with end products as CO$_2$ and H$_2$, and heat is generated through combustion of fuel in the presence of air. The membrane separates and purifies H$_2$ as permeate gas, which is swept away at the downstream side of the membrane. The retentate gas contains mainly CO$_2$ with some unreacted CH$_4$ and water. The SR reaction is endothermic and has higher equilibrium

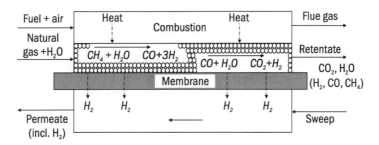

Figure 6   *Typical schematic diagram of hydrogen membrane contactor*

constant at higher temperatures. The WGS reaction is exothermic and has higher equilibrium constant at lower temperatures. A pressure of 36 bar, temperature 400°C–600°C is found to be ideal for HMR operation.

Hydrogen membrane reforming is based on the development of novel ceramic membranes permeable to H_2. The HMR gas power cycle operates with two reactors, one for syn-gas (H_2, CO, and CO_2) production and another for H_2 production from syn-gas. In the first reactor, steam reforming of CH_4 to syn-gas takes place; H_2 that permeates through the membrane is usually combusted with air. The nitrogen and water from this stage may be used as sweep gas in the downstream membrane process or used as diluent for H_2 in the gas turbine for NO_X control. The second reactor is used for separating H_2 and CO_2; H_2 is then used as carbon free fuel for power production. Reforming reactions take place at a pressure of 20 bar and temperature of 700°C–1000°C. Though, the present design of gas turbines allows for a maximum of 50% volume of H_2, the development of burners to handle higher H_2 concentrations is likely to improve economics of pre-combustion power generation [32].

Development of novel ceramic membranes with good H_2 permeablility and high temperature with standability to recover H_2 is an important issue for CO_2 sequestration. Palladium-ceramic composite membrane is such a perm-selective membrane applicable for high temperature gas separations [33]. Electroless plating, controlled autocatalytic deposition of a continuous film on the surface of a substrate by the interaction of a metal salt and a chemical reducing agent, can be used to make thin films of metals, alloys, and composites on both conducting and non-conducting surfaces. Commercially available palladium (Pd) membranes (thicknesses:

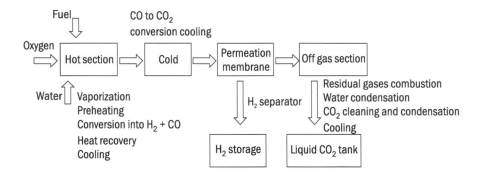

Figure 7 *Typical process flow diagram of Pd-ceramic membrane module*

20–100 μm) are based on self-supporting metals foils. A typical process flow diagram of Pd membrane module is shown in Figure 7.

Major challenges in membrane process involve production of thin layer and low cost membranes. The thinner is the membrane, the higher is the H_2 flux and the lower is the membrane cost. Higher flux also reduces the requirement of number of membranes for a definite amount of load. The thickness of supported Pd membrane layer in the HYSEP® module is as low as 3–9 μm.

Sorption Enhanced Water-Gas-Shift Reaction

Sorption enhanced water-gas-shift (SEWGS) is a potential technology for reduction in CO_2 capture costs in comparison to conventional amine scrubbing [30]. In a typical pre-combustion technology, a mixture of H_2 and CO_2 are produced by high-temperature water-gas-shift reaction (HTS) at around 400°C as discussed above. H_2 is separated from the mixture to be used as fuel and the remaining CO_2 is sequestrated. Since HTS is CO equilibrium limited reaction, the gas is cooled down to 200°C and a low temperature shift reaction is performed to get desired conversion of CO. The gas is cooled amine-based capture of CO_2. It is again heated to around 400°C before feeding into the gas turbine. Thus, the several cooling and heating steps make the process expensive. SEWGS is an alternative process, where HTS is combined with high temperature CO_2 adsorption shifting the equilibrium conversion of CO to the right. Thus, complete CO conversion is achieved and production of H_2 is maximized. In addition, regenerated CO_2 from the adsorbent material is at sufficient purity for direct sequestration. According to Wright *et al.* [34], SEWGS process using a PSA cycle to regenerate the adsorbent can reduce the cost of CO_2 capture. Air products under Phase II of CO_2 capture programme (CCP2) has been working on the development of this technology.

INTEGRATED GASIFICATION COMBINED CYCLE

Integrated gasification combined cycle technology, through the process called gasification, turns low-value fuels such as high-sulphur coal, heavy petroleum residue, biomass or municipal waste into low heating value, high-hydrogen syn-gas. The term "integrated" is used because the gasification unit is integrated in the plant and is also optimized for the plant's combined cycle. Heating value of syn-gas is 4.2–4.6 MJ/ Nm^3, about 3–8 times lower than that of natural gas. Impurities such as sulphur compounds, NH_3, metals, alkalytes, ash, and particulates are removed from the gas before using as fuel for the gas turbine. AC bed is used to remove mercury. Therefore, IGCC power plants have lower emissions [32]. A typical process flow diagram for IGCC technology is shown in Figure 8.

Some of the benefits of IGCC are [35] as follows:
1. Less capital cost, because of low solvent requirement.
2. More than 90% mercury removal.
3. 18% less NO_X than supercritical pulverized coal.
4. 50% less SO_X than supercritical pulverized coal.
5. 42% less particulate matter than supercritical pulverized coal.

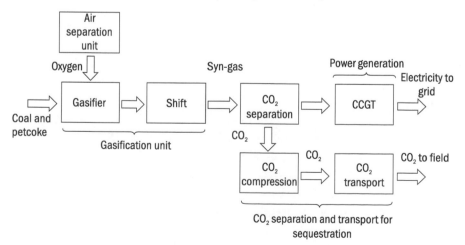

Figure 8 *Typical process flow diagram of IGCC*

6. Inorganic components (metals and minerals) are trapped in an inert and environmentally safe form as ash, which may have use as fertilizer.
7. 30% less water consumption than conventional coal technology
8. Produces half the solid waste of conventional coal technology
9. IGCC technology is most suited for high-sulphur bituminous coals.

GAS PROCESSING PRIOR TO CO_2 CAPTURE

Flue gas from a power plant contains SO_2, NO_X, oxygen, nitrogen, water, and particulates in addition to CO_2. Presence of SO_X can be detrimental to amine-based capture plant operation as it reacts irreversibly with MEA to produce non-reclaimable corrosive salts. For MEA-based processes, a SO_X scrubber is installed before the amine plant, which is usually less expensive than the solvent loss, when the flue gas contains more than 10 ppmv SO_2. About 90%–95% SO_X reductions can be achieved by the limestone or wet lime FGD systems. Further reduction is done by using active alkali metal neutralizing agents such as caustic soda, soda ash, and so on [36]. Similar to SO_2, presence of sulphur trioxide (SO_3) also leads to solvent losses due to formation of heat stable salts as well as corrosive sulphuric acid (H_2SO_4) aerosols in wet scrubbers. Substantial amount of SO_3 is usually recovered through the SO_X scrubber. The limestone/wet lime FGD systems with spray, tray, or packed tower absorbers and the lime spray dryer systems are used for SO_X removal. The removal of SO_3 is usually done in the upstream of the SO_2 scrubber.

'Fuel NO_X' is the NO_X formed by oxidation of nitrogen bound in fuel and 'Thermal NO_X' is the NO_X formed from air at high temperatures associated with combustion. Most of the NO_X formed in combustion process is NO and its production is controlled though control of the peak flame temperature.

NO_X control processes are classified as: Selective Non-Catalytic Reduction (SNCR); Selective Catalytic Reduction (SCR); Combined NO_X/SO_2 processes; Other NO_X only processes. Urea or NH_3 is used to reduce NO_X in the combustion gases to elemental nitrogen in SNCR and SCR. SNCR is sometimes called as thermal reduction as it requires higher temperature. Rate of selective chemical reaction between NO_X and NH_3 to produce nitrogen, and water is increased in SCR, because of catalyst (metal oxide catalysts such as titania vanadia). SCR has the highest NO_X reduction capability (greater than 90%) and is the most widely commercialized post-combustion control technology. It is to be noted that NO_X removal may not be needed in the Indian context as it is within the acceptable limits.

No combined process for the control of both SO_2 and NO_X concentration in the flue gas has yet achieved widespread acceptance although numerous approaches have been proposed. NOXSO process simultaneously removes 90%–95% of the SO_2 and 80% of the NO_X from flue gases.

Tri-Mer TRI-NO_X is a commercial proprietary NO_X control process that works on multistage scrubbing using a solvent that also removes hydrogen chloride, SO_2, chlorine, nitric acid, hydrogen fluoride, and other residual inorganic gases simultaneously with NO_X. Cyclone and bag filters are generally used for removing particulates. Inhibitors are used for protecting equipment corrosion and amine degradation due to the presence of oxygen. Other approaches to remove oxygen are use of expensive alloys, oxygen scavenger, and so on.

GLOBAL R&D AND INDUSTRIAL FACILITIES FOR CO_2 CAPTURE

Some commercial plants across the world use solvent-based capture of CO_2 and also natural gases and syn-gases for reduction/removal of CO_2. Weyburn project, Sleipner project, Lummus scrubber project, Fluor Econamine FGSM scrubber project, and Kansai-Mitsubishi proprietary carbon dioxide recovery (KM CDR) process mostly use MEA for the capture of CO_2. Some of these projects are also working on modification of existing technology and cost reduction. Process performance and economics depends on appropriate selection of the process, solvent, contactor, flue gas flow rate, CO_2 content, amount of impurities for gas processing prior to capture, energy requirement in the stripper, amine recovery, reclaiming, and corrosion.

The end uses of the captured and relatively pure CO_2 are enhanced oil recovery (EOR); production of soda ash and urea; and food processing, freezing, beverage production, and chilling purposes. Some projects are also working on refinement of current technologies. The KM CDR process uses proprietary adsorbent KS-1 to recover about 90% CO_2 from the flue gas. There are installations of this process in a fertilizer plant in Uttar Pradesh, India.

Industries and research organizations such as Cooperative Research Centre for Greenhouse Gas Technologies (CO_2CRC), Australia; industry partners from the UK, USA, Italy, Norway, Netherlands, EC and Canada; and the CCP Project, a joint initiative involving nine of the world's leading energy companies viz., BP, Chevron, ENI, Norsk Hydro, EnCana, Shell, Statoil, Suncor, Energy Texaco Inc. have taken up R&D projects for cost-effective solutions exploring various solvent-based capture systems, membrane and pressure swing adsorption systems, hydrate and distillation systems. In addition to the capture projects, green and clean projects involving 'Advanced Zero Emissions Power Plant' and CO_2-free gas turbine system are also being investigated. In India also, some research institutes including the Indian Institute of Technology (IIT) and Indian Institute of Petroleum (IIP) have taken up research that includes amine-based, membrane-based as well as microbial capture of CO_2.

In addition to the R&D activities for economic and efficient traditional post-combustion CO_2 capture technologies, significant work is being carried out on pre-combustion capture. Project involving co-production of electrical power and H_2 from natural gas with integrated CO_2 capture for zero emissions (FutureGen Industrial Alliance Inc.), the SIMTECHE CO_2 Hydrate Separation Process for separating CO_2 from shifted syngas forming CO_2/H_2O hydrates, palladium-based membrane to reform hydrocarbon fuels to mixtures of H_2 and CO_2, oxy-fuel boiler technology (the ENCAP projects), and microalgal recovery and sequestration of CO_2 from stationary combustion systems by photosynthesis are being investigated.

CONCLUSION

For removal of CO_2 present at low concentrations in the flue gas and/or for high product purity, aqueous primary or secondary alkanolamines are generally employed. The process is energy intensive since huge energy must be supplied in the regeneration of the amines. In view of this, blended amine solvents involving amine with faster reaction and amine with easier regeneration are considered as effective solvent system. Membrane contactor is being projected as effective device for the mass transfer operation of concern. With the renewed interest on adsorption technologies for CO_2 removal/separation, various efficient adsorbents are being synthesized as a promising sorbent for CO_2 and modification of the existing adsorptive separation technologies are also being attempted.

Though, enormous studies are being carried out for technology development, adsorbent synthesis and techno-economic feasibility analysis, commercial installation of adsorptive separation technologies for CO_2 capture in large scale is not in place yet. In addition, various pre-combustion technologies such as membrane reforming, IGCC, etc. are being tried for generating green power. Thus, researches are focused on refinement to current technologies and development of new technologies for both pre- and post-combustion capture of CO_2.

Summary

It is need of the hour to be thorough with the CO_2 capture processes and the advancements as well as merits and demerits of traditional and potential areas such as absorptive, adsorptive, and other emerging technologies. The CO_2 capture technologies can be broadly grouped into physical means and chemical means. MEA is used for chemical separation of CO_2 from flue gases from power plants due to its ability to absorb CO_2 under low partial pressure conditions. The CO_2 removal/separation by physical adsorption has drawn renewed attention, because of development of potential adsorbents and modification of the existing adsorptive separation technologies. Successful commercialization of adsorptive separation technologies for CO_2 capture is yet to be in place. Advancement in HMR, a combination of SR and WGS reaction modelled into a single unit, is also referred to as MWGSR. SEWGS is used for shifting the equilibrium conversion of CO for its complete conversion into CO_2 and H_2 production maximization. Integrated IGCC turns high-sulphur coal, heavy petroleum residue, biomass or municipal waste into low heating value, high-hydrogen synthesis gas and then removes impurities before it is used as a primary fuel for a gas turbine. Growing concerns over climate change have led to a strong emphasis on the R&D of high-efficiency and economic CO_2 capture technologies.

REFERENCES

1. Aaron, D. and C. Tsouris. 2005. Separation of CO_2 from flue gas: A review. *Sep. Sci. Technol.* 40:321–348

2. Bottoms, R. R. 1930. Process of separating acidic gases. Girdler Corp. US Patent 1783901

3. Kohl, A. L. and R. B. Nielsen. 1997. *Gas Purification*, 5th edn. Houston: Gulf Publishing Company

4. Mandal, B. 2004. *Ph.D. Dissertation.* IIT Kharagpur
5. Mandal, B. and S. S. Bandyopadhyay. 2005. Simultaneous absorption of carbon dioxide and hydrogen sulfide into aqueous blends of 2-amino-2-methyl- 1-propanol and diethanol amine. *Chemical Engineering Science* 60:6438–6451
6. Chakma, A., A. K. Mehrotra, and B. Nielsen. 1995. Comparison of chemical solvents for mitigating CO$_2$ emissions from coal-fired power plants. *Heat Recovery Systems* and *CHP* 15:231–240
7. Ruthven, D. M. 1984. *Principles of Adsorption and Adsorption Processes.* New York: Wiley
8. Ruthven, D. M., S. Farooq, and K. S. Knaebel. 1990. *Pressure Swing Adsorption.* New York: VCH Publishers Inc.
9. Knöfel, C., J. Descarpentries, A. Benzaouia, V. Zelenák, S. Mornet, P. L. Llewellyn, and V. Hornebecq. 2007. Functionalised micro-/mesoporous silica for the adsorption of carbon dioxide. *Micropor. Mesopor. Mater* 99:79–85
10. Wei, J., J. Shi, H. Pan, W. Zhao, Q. Ye, and Y. Shi. 2008. Adsorption of carbon dioxide on organically functionalized SBA-16. *Microporous and Mesoporous Materials* 116:394–399
11. Khomane, B. R., B. K. Sharma, S. Saha, and B. D. Kulkarni. 2006. Reverse micro emulsion mediated sol-gel synthesis of lithium silicate nanoparticles under ambient conditions: Scope for CO$_2$ sequestration. *Chemical Engineering Science* 61:3415–3418
12. Manjare, S. D. 2004. Studies on adsorption of ethyl acetate vapour in molecular sieves. *Ph. D. Thesis,* Birla Institute of Technology and Science, Pilani, India
13. Alvim-Ferraz, M. C. M., and C. M. Todo-Bom. 2003. Impregnated active carbons to control atmospheric emissions. *Journal of Colloid and Interface Science* 259:133–138
14. Arenillas, A., F. Rubiera, J. B. Parra, C. O. Ania, and J. J. Pis. 2005. A comparison of different methods for predicting coal devolatilisation kinetics. *Applied Surface Science* 252:619–624
15. Manocha, S. M. 2003. Porous carbons. *Sadhana* 28:335–348
16. Guo, B., L. Chang, and K. Xie. 2006. Adsorption of carbon dioxide on activated carbon. *Journal of Natural Gas Chemistry* 15:223–229
17. Pires, J., M. Bestilleiro, M. Pinto, and A. Gil. 2008. Selective adsorption of carbon dioxide, methane and ethane by porous clays heterostructures. *Separation and Purification Technology* 61:161–167
18. Cavenati, S., C. A. Grande, and A. E. Rodrigues. 2006. Separation CH$_4$/CO$_2$/N$_2$ mixtures by layered pressure swing adsorption for upgrade of natural gases. *Chemical Engineering Science* 61:3893–3906

19. Ghezini, R., M. Sassi, and A. Bengueddach. 2008. Adsorption of carbon dioxide at high pressure over H-ZSM-5 type zeolite. Micropore volume determinations by using the Dubinin–Raduskevich equation and the "*t*-plot" method. *Microporous and Mesoporous Materials* 113:370–377

20. Zhang, J., P. A. Webley, and P. Xiao. 2008. Effect of process parameters on power requirements of vacuum swing adsorption technology for CO_2 capture from flue gas. *Energy Conversion and Management* 49:346–356

21. Park, J. H., H. T. Beum, J. N. Kim, and S. H. Cho. 2002. Numerical analysis on the power consumption of the PSA process for recovering CO_2 from flue gas. *Ind. Eng. Chem. Res.* 41:4122–4131

22. Diagne, D., M. Goto, and T. Hirose. 1995. Parametric study on CO_2 separation and recovery by dual reflux PSA process. *Ind. Eng. Chem. Res.* 34:3083–3089

23. Chue, K. T., J. N. Kim, Y. J. Yoo, S. H. Cho, and R. T. Yang. 1995. Comparison of activated carbon and zeolite 13X for CO_2 recovery from flue gas by pressure swing adsorption. *Ind. Eng. Chem. Res.* 34:591–598

24. Ishibashi, M., H. Ota, N. Akutsu, S. Umeda, M. Tajika, J. Izumi, A. Yasutake, T. Kabata, and Y. Kageyama. 1996. Technology for removing carbon dioxide from power plant flue gas by the physical adsorption method. *Energ. Convers. Manage.* 37:929–933

25. Sarofim, A. F. 2007. Oxy-fuel combustion: Progress and remaining issues. *IEAGHG International Oxy-Combustion Research Network, 2nd Workshop,* 25–26 January, Windsor, CT, USA

26. Liu, H. and K. Okazaki. 2003. Simultaneous easy CO_2 recovery and drastic reduction of SOx and NOx in O_2/CO_2 coal combustion with heat recirculation. *Fuel* 82:1427

27. Corbella, B. M., L. F. de Diego, F. García-Labiano, J. Adánez, and J. M. Palacios. 2006. Characterization and performance in a multicycle test in a fixed-bed reactor of silica-supported copper oxide as oxygen carrier for chemical-looping combustion of methane. *Energy & Fuels* 20:148–154

28. Jerndal, E., T. Mattisson, and A. Lyngfelt. 2006. Thermal analysis of chemical-looping combustion. *Trans I Chem E, Part A Chem. Eng. Res. and Des.* 84:795–806

29. McGlashan, N. R. 2008. Chemical looping combustion–a thermodynamic study. *Proc. I. Mech. E. Part C: J. Mech. Eng. Sci.* 222:1005–1019

30. Metz, Bert, Ogunlade Davidson, Heleen de Coninck, Manuela Loos, and Leo Meyer (eds). 2005. IPCC special report on carbon dioxide capture and storage, p. 129. *Prepared by Working Group III of the Intergovernmental Panel on Climate Change.* New York: Cambridge University Press

31. Abanades, J. C., J. Edward, A. J. Wang, and J. E. Oakey. 2005. Cost structure of a postcombustion CO2 capture using CaO. *Environmental Science and Technology* 39:2861–2866

32. Miraccaa, I., K. I. Åsenb, J. Assinkc, C. Coulterd, L. Currane, C. Lowef, G. T. Moureg, and S. Schlasnerh. 2009. The CO_2 capture project (CCP): Results from phase II (2004–2009). *Energy Procedia* 1(1): 55–62

33. Ilias, S., F. G. King, T-F Fan, and S. Roy. 1996. *Advanced Coal-Fired Power Systems '96 Review Meeting*, 16–18 July

34. Wright, A., V. White, J. Hufton, E. van Selow, and M. Hinderink. 2009. *Energy Procedia* 1:707–714

35. USEPA. 2006. Global mitigation of non-CO_2 greenhouse gases (Report No. EPA 430-R-06-005) *U. S. Environmental Protection Agency*

36. Brady, J. D. 1986. Flue gas scrubbing process for sulphur dioxide and particulate emission preceding carbon dioxide absorption. *Conference Paper No. 31d, American Institute of Chemical Engineers Spring National Meeting* 6–10 April, New Orleans, Louisiana, USA

CHAPTER 6

CO_2 Recovery From Power Plants by Adsorption: R & D and Technology

Anshu Nanoti, Madhukar O. Garg, Aamir Hanif, Soumen Dasgupta, Swapnil Divekar, and Aarti

CSIR-Indian Institute of Petroleum, Dehradun-248005

E-mail: anshu@iip.res.in

INTRODUCTION

Rising levels of carbon dioxide (CO_2) in the atmosphere due to burning of fossil fuels for power generation have been recognized to be the main contributor of global warming and associated climate change phenomenon. There are three generic methodologies that can be used for CO_2 capture from a power plant [1]. These are post-combustion CO_2 capture, pre-combustion CO_2 capture, and oxy-fuel combustion capture. Each methodology has its own merits and demerits. A number of adsorbent materials such as zeolites and activated carbons (ACs) as well as hydrotalcites for high temperature applications have been investigated [2–3].

In pre-combustion approach, gasification of feedstock (coal) produces a hot multi-component gas stream containing acidic gases such as hydrogen sulphide (H_2S) and CO_2 along with hydrogen (H_2). Conventional solvent-based technology for removal of acid gases operates at low temperatures; hence, the gas clean-up train requires cyclic heating and cooling steps to produce clean H_2. These temperature swings lead to overall lower thermal efficiency of the process [4]. Developing an alternative adsorption-based technology using adsorbents, which can work at higher temperatures for CO_2 and H_2S after removal, will be a step change towards increasing the thermal efficiency of the process.

Post-combustion is an end-of-pipe treatment of flue gas for removal of CO_2 present prior to discharge through the stack [5]. The CO_2 levels are generally in the range of 5%–15% depending on the type of fuel undergoing combustion and it must be separated from the mixture containing nitrogen (N_2), oxygen, moisture, and oxides of sulphur (SO_x)/ oxides of nitrogen (NO_x) as impurities. In the post-combustion approach, solid adsorbents such as zeolites and ACs can be used to recover CO_2 from the flue gas mixture using pressure swing adsorption (PSA) technique. Several adsorbent materials have been investigated for CO_2 recovery by both PSA and vacuum swing adsorption (VSA). The general consensus is that zeolite 13X materials perform better than ACs or silica gels [6]. Both capacities and selectivities for separation of CO_2/N_2 mixtures (representative of flue gases from power plants) are superior. However, power requirement during regeneration can be high and for this reason, there is a large scope for developing new adsorbents, which will show better selectivity and regenerability.

Metal organic framework (MOF) is a new class of adsorbents [7] attracting interest for selective CO_2 separation. These are materials in which metal ions or clusters are connected via organic linkages to form highly porous network structures. Several MOFs have been proposed as adsorbents for CO_2 recovery. However, most studies that have been reported so far on CO_2 adsorption on MOFs are limited to equilibrium isotherms and diffusion measurements with pure components. Not much data are available on adsorption processes, such as pressure–volume swing adsorption (PVSA). The focus of this chapter is to project the research carried out at Council of Scientific and Industrial Research-Indian Institute of Petroleum (CSIR-IIP) in adsorption-based CO_2 capture technologies. The study includes CO_2 removal from two types of gas feeds representing both pre-combustion and post-combustion process streams. The materials used are zeolite, MOF, and hydrotalcite-based adsorbents.

CO₂ CAPTURE FROM POST-COMBUSTION FLUE GAS

At CSIR-IIP, a collaborative project under the aegis of National Thermal Power Corporation (NTPC) Ltd was carried out to recover CO_2 in high purity from coal-fired power plant flue gases. A novel VSA-based process was developed using zeolite-based adsorbent materials at moderate temperatures and at low partial pressure for CO_2 capture from power plant flue gas [8]. In a typical VSA cycle, desorption step is carried out under evacuation to sub-atmospheric pressures.

Conventional PSA-based processes are extensively used in petroleum refineries for H_2 production and are based on well-known Skarstrom cycle, which is originally designed to produce weakly adsorbed components such as H_2 in high purity, but is not suitable to produce strongly adsorbed component such as CO_2 in high purity [9]. However, when CO_2 is the desired product to be recovered in high purity then there is a need to develop new type of cycles. Recent studies have reported that good CO_2 enrichment can be achieved by judiciously incorporating a heavy reflux step, which involves partial recycling of a CO_2 enriched gas stream to an adsorption column for saturating adsorbent bed with CO_2. This in turn improves the purity of the recovered CO_2 product from the column [10]. Complex three- to eight-bed PVSA cycles employing this heavy reflux concept were proposed for CO_2–N_2 separation using AC, 13X zeolite, and promoted hydrotalcites as adsorbents. In some of them, very high purity of CO_2 (>99 mole%) could be obtained at moderate recovery [11].

The novel eight-bed PVSA cycle designed and tested at CSIR-IIP in a custom built automated three column unit is shown in Figure 1. The typical step sequences of the PVSA cycle are given in Table 1.

The flue gas separation performance of the cycle was optimized with respect to process parameters such as feed flow rate, adsorption time, evacuation time, extent of CO_2 rinse, and so on. The cycle performance was optimized at 2 bar adsorption pressure and at 55°C temperature. Under optimized conditions, >90 mole% CO_2 purity was achieved at around 80% recovery with a good productivity of 2.68 NL/kg/min from CO_2 levels of 13 mole% in flue gas.

Figure 1 *Three column lab scale PVSA unit at CSIR-IIP, Dehradun*

Table 1 : Sequencing of process steps in three column configuration for CO_2 recovery from flue gas

Cycle steps	Column 1	Column 2	Column 3
1	Feed pressurization	Pressurization with part of CO_2 enriched evacuation product of Column 3	Evacuation
2	Adsorption	Co-current depressurization	
		Rinse with part of CO_2 enriched evacuation product from Column 3	Evacuation with adsorption product purge
3	Co-current depressurization		Pressure equalization
4	Pressurization with part of CO_2 enriched evacuation product of Column 2	Evacuation	Feed pressurization
5	Co-current depressurization		Adsorption
6	Rinse with part of CO_2 enriched evacuation product from Column 2	Evacuation with adsorption product purge	
		Pressure equalization	Co-current depressurization
7	Evacuation	Feed pressurization	Pressurization with part of CO_2 enriched evacuation product of Column 1
			Co-current depressurization
8	Evacuation with adsorption product purge	Adsorption	Rinse with part of CO_2 enriched evacuation product from Column 1

Reappraisal of Skarstrom cycle for CO_2 recovery [12] suggested that contrary to conventional wisdom, the flue gas separation performance of a potassium exchanged X zeolite can be better with Skarstrom-type VSA cycle than heavy reflux cycle. The key parameter here seems to be easy regenerability of the adsorbent, which led to a high purity CO_2 stream during the regeneration of the bed under evacuation.

The CO_2 purity achieved with Skarstrom cycle was 87 mole% at a recovery 75% from a feed consisting of 15 mole% CO_2 in N_2. The CO_2 purity and recovery level obtained with the Skarstrom cycle was 6%

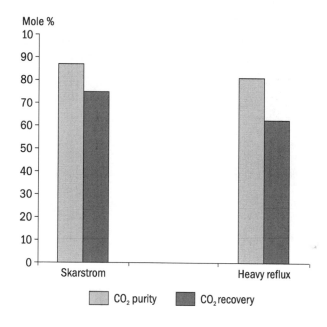

Figure 2 *Comparison of Skarstrom and heavy reflux VSA cycle performance for flue gas separation with KX zeolite adsorbent*

and 19% more respectively, compared to the heavy reflux cycle under same operating conditions of adsorption pressure, temperature, feed flow rate, feed composition, adsorption, and evacuation time. The results are shown in Figure 2.

APPLICATION OF METAL ORGANIC FRAMEWORKS IN CO$_2$ CAPTURE FROM FLUE GAS

Metal organic frameworks are attracting interest for selective CO$_2$ separation as a new class of adsorbents. These are materials in which metal ions or clusters are connected via organic linkages to form highly porous network structures. Several MOFs have been proposed as adsorbents for CO$_2$ recovery. However, studies reported on CO$_2$ adsorption on MOFs have been limited mostly to equilibrium isotherms and diffusion measurements with pure components. One important aspect to be considered for MOF's application in process is its formulation. Synthesized MOF materials being fluffy powders, they are not suitable for industrial processes. Formulation into well-shaped particles is therefore necessary to come up one step closer to industrial applications. Little attention has been paid on this important aspect despite explosive growth in MOFs research in recent years.

Feasibility of a VSA process for the separation of CO_2 from simulated flue gas by MOF-based adsorbents has been studied at CSIR-IIP in collaboration with SINTEF Materials and Chemistry, Oslo, Norway.* The major focus of IIP-SINTEF collaboration is on effective formulation of MOF adsorbents in order to make them suitable for column adsorption studies specifically for VSA-based CO_2 separation process from the flue gas. After a thorough screening of a range of MOF materials based on equilibrium CO_2, N_2 uptake, isosteric heat of adsorption, surface area, stability, ease of handling, and so on, two MOFs, namely CPO-27-Ni and UiO-66, were shortlisted for detailed evaluation. Both materials gave significant CO_2 adsorption at 10 kPa CO_2 pressure (approximately 4.2 and 2.5 mmol/g at 298K, for CPO-27-Ni and UiO-66, respectively) [13–14].

The VSA cycle studies were carried out in a single column microadsorber unit at CSIR-IIP (Figure 3) with MOFs obtained from SINTEF as well as commercially available MOFs (CuBTC) from Sigma Aldrich. The results were compared with 13X zeolite [15].

Figures 4–5 show some of the representative results for single column VSA studies. Results indicate the performance of zeolite in terms of CO_2 purity obtained is better than MOF material. However, CO_2 recoveries obtained with MOF material were consistently more than obtained with zeolite, which clearly indicates better regenerability of metal organic

Figure 3   *Single column microadsorber unit*

* *Funded by Norwegian Research Council (NRC)*

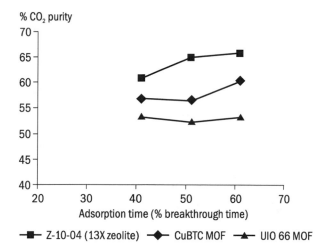

Figure 4 *Effect of adsorption time on CO₂ purity: a comparison between zeolite and MOF studied*

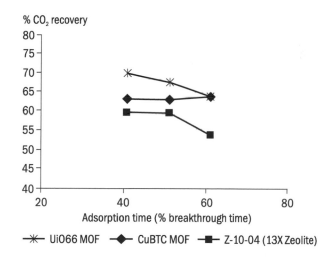

Figure 5 *Effect of adsorption time on CO₂ recovery: a comparison between zeolite and MOF studied*

frameworks. Hence, for low-pressure applications such as CO₂ removal from flue gas, MOF adsorbents could be a preferred choice.

HIGH TEMPERATURE CO₂ CAPTURE UNDER WATER GAS SHIFT REACTOR CONDITIONS

Gas clean-up technologies, presently available for removal of CO₂ in pre-combustion process, are based on chemical or physical absorption and operate at ambient/sub-ambient temperatures. Implementation of these technologies on coal gas leads to high thermal efficiency loss, in the case of both sweet water gas shift and sour water gas shift due to the requirement of cooling and re-heating of the syn-gas.. Therefore, it is advantageous to effect CO_2 removal in hot gas conditions in terms of increased energy efficiency and considerable research and development (R&D) efforts are being made towards this end. The R&D trend is now towards CO_2 sorption at temperatures in the range 300°C–600°C. The high-temperature CO_2 sorbents mostly studied are calcium oxides [16], lithium zirconates [17], metal oxides [18], and hydrotalcites [19]. Hydrotalcites are a class of anionic clays that have been evolved as the best choice due to most of the desirable properties such as lower energy of regeneration, retention of capacity after multiple cycles, and suitable kinetics of CO_2 adsorption. The hydrotalcites, however, have lower CO_2 capture capacity as compared to other known high-temperature sorbents. Continuous research efforts have shown that the CO_2 capture capacity can be improved by various manipulations in its chemical composition.

For high-temperature capture of CO_2 under sour water gas shift reactor conditions (temperature: 300°C–400°C; pressure: 20 Bar), pressure–temperature swing adsorption (PTSA) process has been developed in research mode. The sorbent used in the process is a synthetic hydrotalcite code named HTA (CP)-1. The sorbent was synthesized by co-precipitation from an aqueous mixture of aluminium nitrate and magnesium nitrate. An improvement in CO_2 sorption capacity was observed following potassium carbonate (K_2CO_3) impregnation (Figure 6).

The adsorbent is found to be regenerable under inert purge with or without evacuation at ~450°C, which indicates requirement of a small temperature swing of 50°C–100°C for adsorbent regeneration. Also, no appreciable loss in CO_2 breakthrough capacity was observed over multiple cycles of adsorption–regeneration (Figure 7).

Result shows that high-temperature CO_2 capture capacity of the hydrotalcite sorbent studied at our laboratory is comparable to the best reported values for similar class of high-temperature sorbent [20]. We have observed a temperature-dependent promotional effect of K_2CO_3 impregnation on CO_2 uptake, the highest CO_2 uptake (~2 mmol CO_2/g sorbent @ 10 bar CO_2 pressure) being in the temperature range 350°C–400°C for the promoted adsorbent. The sorbent is regenerable

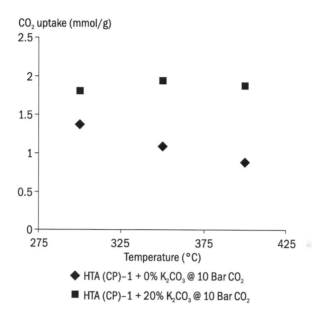

Figure 6 *Promotional effect of K₂CO₃ impregnation on CO₂ loading of a hydrotalcite at high temperatures*

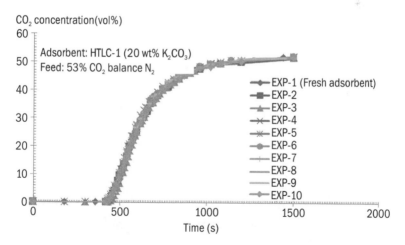

Figure 7 *Successive CO₂ breakthrough experiments at 400°C and 12 bar pressure*

under inert purge with or without evacuation at ~450°C, which indicates that a smaller temperature swing of 50°C–100°C will be required for adsorbent regeneration. No appreciable loss in breakthrough capacity was observed over multiple cycles of adsorption–regeneration.

CONCLUSION

The research programme on CO_2 capture at CSIR-IIP has been carried out in collaboration with both national laboratories and international organizations. This programme aims to develop adsorption-based processes for CO_2 recovery in high purity both from post-combustion and from pre-combustion power plant flue gases using advanced adsorbent materials as well as to develop novel PVSA cycles under operating conditions representing actual process streams. Results of different adsorbent materials based on zeolites, formulated MOFs, and mixed oxide evaluated under different PVSA cycles for CO_2 recovery from flue gases are presented here with technology challenges in them.

Summary

There are three generic methodologies that can be used for CO_2 capture from a power plant. These are post-combustion CO_2 capture, pre-combustion CO_2 capture, and oxy-fuel combustion capture. Each methodology has its own merits and demerits. In this chapter, research on adsorption-based capture technologies carried out in CSIR-IIP were presented. Chemical kinetics and material needs for zeolites, MOFs, and hydrotalcite-based adsorbents in pre-combustion and post-combustion capture were described with technology challenges.

REFERENCES

1. Yang, H., Z. Xu, M. Fan, R. Gupta, R. B. Slimane, A. E. Bland, and I. Wright. 2008. Progress in carbon dioxide separation and capture: A review. *Journal of Environmental Sciences* 20:14–27

2. Yavuz, C. T., B. D. Shinall, A. V. Iretskii, M. G. White, M. Atilhan, P. C. Ford, G. D. Stucky, and T. Golden. 2009. Markedly improved CO_2 capture efficiency and stability of Gallium substituted hydrotalcites at elevated temperatures. *Chemistry of Materials* 21:3473–3475

3. Choi, V., J. H. Drese, and C. W. Jones. 2009. Adsorbent materials for carbon dioxide capture from large anthropogenic point sources. *ChemSusChem* 2:796–854

4. Ciferno, J., S. Chen, and W. C. Yang. 2008. Analyses of hot/warm CO_2 capture for IGCC processes. *AIChE Spring National Meeting,* New Orleans, Los Angeles

5. Manovic, V. and E. J. Anthony. 2010. Carbonation of CaO-base sorbents enhanced by steam addition. *Industrial & Engineering Chemistry Research* 49:9105–9110

6. Chue, K. T., J. N. Kim, Y. J. Yoo, S. H. Cho, and R. T. Yang. 1995. Comparison of activated carbon and zeolite 13X for CO_2 recovery from flue gas by pressure swing adsorption. *Industrial & Engineering Chemistry Research* 34:591–598

7. Rowsell, J. L. C. and O. M. Yaghi. 2004. Metal-organic frameworks: A new class of porous materials. *Microporous and Mesoporous Materials* 73:3–14

8. Goswami, A. N., A. Nanoti, P. Gupta, S. Dasgupta, M. O. Garg, H. P. Dulhadinomal, R. Mukhopadhyay, and R. Swami. 2011. Improved VSA process for CO_2 recovery from power plant flue gas. Indian Patent Application 2654 DEL

9. Agarwal, A., L. T. Bieglar, and S. E. Zitney. 2010. A superstructure-based optimal synthesis of PSA cycles for post-combustion CO_2 capture. *AIChE Journal* 56:1813–1828

10. Na, B., H. Lee, K. Koo, and H. K. Song. 2002. Effect of rinse and recycle methods on the pressure swing adsorption process to recover CO_2 from power plant flue gas using activated carbon. *Industrial & Engineering Chemistry Research* 41:5498–5503

11. Ebner, A. D. and J. A. Ritter. 2009. State of the art adsorption and membrane separation processes for CO_2 production from CO_2 emitting industries. *Separation Science and Technology* 44:1273–1421

12. Nanoti, A., S. Dasgupta, Aarti, N. Biswas, A. N. Goswami, M. O. Garg, S. Divekar, and C. Pendem. 2012. Reappraisal of the Skarstrom cycle for CO_2 recovery from flue gas streams: New results with potassium-exchanged zeolite adsorbent. *Industrial & Engineering Chemistry Research* 51:13765–13772

13. Andersen, A., S. Divekar, S. Dasgupta, J. H. Cavka, Aarti, A. Nanoti, A. Spjelkavik, A. N. Goswami, M. O. Garg, and R. Blom. 2012. On the development of Vacuum Swing Adsorption (VSA) technology for post-combustion CO_2 capture. *Eleventh Green House Gas Technology Conference*, Kyoto, Japan

14. Arya, A., S. Dasgupta, S. Divekar, A. Nanoti, A. N. Goswami, M. O. Garg, A. Andersen, J. H. Cavka, and R. Blom. 2012. Single column VSA studies for CO_2 recovery using metal organic framework adsorbent: Comparison with commercial zeolites. *Twelfth AIChE Annual Meeting*, Pittsburgh, USA

15. Dasgupta, S., N. Biswas, Aarti, N. G. Gode, S. Divekar, A. Nanoti, and A. N. Goswami. 2012. CO_2 recovery from mixtures with nitrogen in a vacuum swing adsorber using metal organic framework adsorbent: A comparative study. *International Journal of Greenhouse Gas Control* 7:225–229

16. Han, C. and D. P. Harrison. 1994. Simultaneous shift reaction and carbon dioxide separation for direct production of hydrogen. *Chemical Engineering Science* 49:5875–5883

17. Fernndez, E. O., M. Rønning, X. Yu, T. Grande, and D. Chen. 2008. Compositional effects of nanocrystalline lithium zirconate on its CO_2 capture properties. *Industrial & Engineering Chemistry Research* 47:434–442

18. Xiao, G., R. Singh, A. Chaffee, and P. Webley. 2011. Advanced adsorbents based on MgO and K_2CO_3 for capture of CO_2 at elevated temperatures. *International Journal of Greenhouse Gas Control* 5:634–639

19. Selow, E. R., P. D. Cobden, P. A. Verbraeken, J. R. Hufton, and R. W. Brink. 2009. Carbon capture by sorption enhanced water-gas shift reaction process using hydrotalcite based material. *Industrial & Engineering Chemistry Research* 48:4184–4193

20. http://cordis.europa.eu/documents/documentlibrary/118296641EN6.pdf

CHAPTER 7

Geological Sequestration of CO_2 in Saline Aquifers – an Indian Perspective

A. K. Bhandari

Senior Advisor, Federation of Indian Mineral Industries, FIMI House,
B-311, Okhla Industrial Area, Phase I, New Delhi-110020
E-mail: bhandari_ak@yahoo.com

INTRODUCTION

India has set a goal of sustained economic growth rate of 8%–9% per annum. To achieve this, our energy needs will grow rapidly in future. Ambitious plans are on way to harness hydroelectric potential and fully develop non-conventional energy sources (geothermal, nuclear, and wind). In the coming decades, as fossil fuels, particularly coal, would inevitably be the mainstay for energy generation in India, carbon dioxide (CO_2) emissions are likely to increase exponentially. It is believed that increasing concentration of CO_2 in the atmosphere has the potential to force changes towards warmer global climate. Financial institutions warn that the increasing frequency of severe climate coupled with social trends may result in increased cost related to disaster relief and insurance. It may burden the taxpayers and industries alike.

To contain and ultimately reduce the CO_2 concentrations, geological storage of excess CO_2 has grown from a concept to a potentially important technology driven mitigation option. The information and experience gained from the injection and/or storage of CO_2 from a large number of existing enhanced oil recovery (EOR) and acid gas projects as well as from other international projects such as Sleipner, Weyburn, and In Salah indicate that it is feasible to store CO_2 in geological formations as a CO_2 mitigation option. It is understood that saline aquifers are generally unused and of no economic significance, so there is no value addition from CO_2 storage, but can be a vast reservoir. Little is known about the presence of deep saline aquifers in different geological formations in India. The main constraint is, therefore, lack of detailed knowledge about potential storage sites.

GEOLOGICAL STORAGE OF CO_2

Capturing CO_2 from the major stationary sources and its storage into deep geological formations is considered as a potential mitigation option. Geological storage of CO_2 can be undertaken in a variety of geological settings in sedimentary basins. The options for CO_2 are as follows:

- Depleted oil and gas reservoirs
- Use of CO_2 in EOR
- Deep unmineable coal seams/enhanced coalbed methane (ECBM) recovery
- Oceans
- Deep unused saline water-saturated formations
- Other geological media are basalts, shales, and cavities

Several research programmes on the geological storage of CO_2 are underway in many parts of the world. In little over a decade, the concept of geological storage of CO_2 has grown and it appears that it could make deep cuts to atmospheric CO_2 emissions. Each of the geological storage options (Figure 1) other than saline aquifers is briefly discussed in the ensuing sections.

Depleted Oil and Gas Fields

Depleted oil and gas reservoirs are prime candidates for CO_2 storage as they provide value addition proposition for recovery of enhanced oil and gas. CO_2 can also be stored permanently in empty space of a depleted

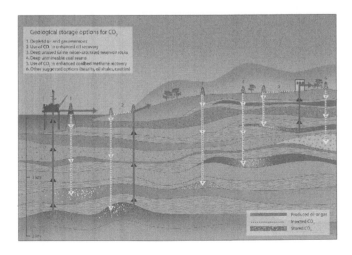

Figure 1 *Options for storing CO_2 in deep underground geological formations*

gas field. The geological structure and physical properties of most oil and gas fields have been extensively studied and characterized.

Enhanced Oil Recovery

Enhanced oil recovery through CO_2 flooding (by injection) is a mature technology and offers potential economic gain from incremental oil production. An incremental oil recovery varying from 7% to 23% has been achieved. For enhanced CO_2 storage in EOR operations, oil reservoirs may need to meet additional criteria.

Coal Seams/ECBM

Abandoned or uneconomic coal seams are other potential CO_2 storage options. Coal contains fractures (cleats) that impart some permeability to the system. It has a higher affinity to adsorb gaseous CO_2 as compared to methane (CH_4). If CO_2 is injected into coal seams, it can displace CH_4, thereby enhancing coalbed methane (CBM) recovery. CO_2 ECBM has the potential to increase the amount of produced CH_4 up to nearly 90%. Coal permeability is one of the several determining factors in the selection of a storage site.

Oceans

As early as 1977, CO_2 storage in oceans had been proposed as a possible disposal medium. Although, sea water itself contains large quantities of CO_2 and the system is in dynamic exchange with atmospheric CO_2 depending on temperature and salinity, it can act as a sink or a source. There are various options proposed for CO_2 storage in oceans at different depths. However, because of the uncertainty of the storage and associated legal issues, ocean storage is unlikely to be promoted as a storage option.

Other Geological Media

Besides saline aquifer, which is discussed in more detail, other geological media include basalts, oil or gas shale, salt caverns, and abandoned mines having a limited potential.

DEEP SALINE AQUIFERS

Deep saline formations are believed to have by far the largest capacity for CO_2 storage worldwide at a relatively little cost and are much more widespread than other options. Sedimentary rocks such as sandstone and limestone have many small spaces or pores that can be filled with water, trapped by an overlying layer of non-porous rock. There are a number of places where deep salt-water reservoirs have been used as buffer stores for natural gas giving confidence that CO_2 could be stored safely for thousands of years in carefully selected reservoirs. If CO_2 is injected into these deep reservoirs, some will dissolve in the saline water and become widely dispersed in the reservoir. CO_2 can also react with the minerals within the reservoir and remain fixed for eternity. The most suitable reservoirs are those at depths greater than 800 m, as CO_2 will behave more like a liquid than a gas, enabling much more quantity to be stored.

Why Saline Aquifers?

Saline formations occur in almost all sedimentary basins throughout the world, both onshore and on continental shelves. Though sequestration of CO_2 in deep saline formations does not result in any value added product, it has other advantages as follows:

(i)   The estimated storage capacity of the saline formations is large enough to make them viable for any long-term solution.

(ii) Saline formations underlie many parts of the world and can be in the proximity of the stationary polluting sources thereby reducing the cost of infrastructure.

(iii) It can help in achieving near zero emissions for the existing power plants and industrial units.

(iv) The fact that CO_2 has been naturally stored for geological time scales enhances the creditability of the storage options.

(v)   Huge thickness of impervious (clay/sandstone) cap rock ensures that residence times are long and accounting for volume sequestered is straightforward.

(vi) Scenarios for negative impacts and unintended damages are limited.

(vii) Usually due to their high saline proportions and depth, they cannot be technically and economically exploited for surface uses.

MECHANISMS OF CO_2 TRAPPING IN SALINE AQUIFERS

For any storage site, the depth of the aquifer and a cap rock are critical factors. Usually only aquifers 800 m below sea level are considered for CO_2 storage. At temperatures and pressures in the sub-surface of around 600 m to 800 m, CO_2 exists in its dense phase as a liquid and occupies much less pore volume than in its gaseous phase. 1 tonne of CO_2 occupies 509 m³ at surface conditions of 0°C and 1 bar. The same amount of CO_2 would occupy only 1.39 m³ at 1000 m sub-surface conditions of 35°C and 102 bar. In addition to a reservoir rock, an overlying "cap rock" that is impermeable to the passage of CO_2 is required. When CO_2 is injected into a reservoir, it is more buoyant than the reservoir fluid in the pore spaces and prevents the CO_2 migrating vertically and so the CO_2 becomes trapped at the top of the reservoir underneath the cap rock. The cap rock provides the main trapping mechanism for the long-term security of storage. Cap rocks are usually shales, mudstones, or evaporite layers. The cap rock should ideally be unfaulted, as faults would provide migration pathways for the CO_2 out of the reservoir.

There are multiple mechanisms for storage including physical trapping beneath low-permeability cap rock, dissolution, and mineralization. Storage of CO_2 in saline aquifers can be in both "confined" and "unconfined" aquifers (Figure 2). Storage in confined aquifers relies on tapping of the buoyant CO_2 by structural and stratigraphic trapping. Physical or geochemical trapping occurs in unconfined aquifers.

Physical Trapping: Stratigraphic and Structural

Initially, physical trapping of CO_2 below low-permeability seals (cap rocks), such as shale or salt beds, is the principal means to store CO_2 in geological formations. Sedimentary basins have such closed,

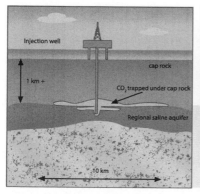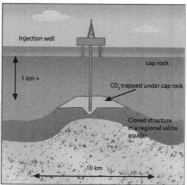

Figure 2 *Conceptual diagram of storage in (a) confined and (b) unconfined aquifers*

physically bound traps or structures, which are occupied mainly by saline water. Structural traps include stratigraphic traps and those formed by folded or fractured rocks.

Physical Trapping: Hydrodynamic

Hydrodynamic trapping can occur in saline formations that do not have a closed trap.

Storage in unconfined aquifers involves injection of CO_2 into large regional aquifers with no specific large structural or stratigraphic closures as a target. Once CO_2 has been injected, it migrates upwards along the most permeable pathway until it encounters the impermeable cap rock, which inhibits further vertical movement. The CO_2 then migrates largely laterally, along the cap rock–reservoir boundary and following the most permeable pathways in the uppermost parts of the aquifer, it will fill small domes and undulations in the reservoir underneath the cap rock, effectively trapping a proportion of the injected CO_2. The CO_2 will then migrate laterally. In unconfined aquifers, it is likely that over time the CO_2 will become distributed over a large area and in low concentrations.

Geochemical Trapping

Carbon dioxide in the sub-surface can undergo a sequence of geochemical interactions with the rock and formation water. Some of the CO_2 dissolves in the formation water through solubility trapping. Some fraction may be converted to stable carbonate minerals (mineral trapping), the most permanent form of geological storage. Mineral trapping is believed to be comparatively slow, potentially taking a thousand years or longer. Nevertheless, the permanence of mineral storage combined with the potentially large storage capacity present in some geological settings makes this a desirable feature of long-term storage.

SPECIFIC ISSUES CONNECTED WITH STORAGE IN SALINE AQUIFERS

To secure long term and safe storage, the following studies are needed.

(i) **Reservoir properties and modelling** In contrast to storage in abandoned oil or gas fields, where there is likely to be abundant core, geophysical logs, pressure measurements, both of the reservoir and of the overburden and seismic data, all of which can help to characterize the reservoir, saline aquifers are not likely to be as well-studied. Therefore, reservoir characterization and the building of a model to test migration paths will most likely require the acquisition of such data, thereby increasing cost significantly.

(ii) **Cap rock integrity** In hydrocarbon fields, the existence of a seal is demonstrated by the very existence of the field. In the case of aquifers, determination of cap rock properties by testing and modelling is likely to be required.

(iii) **Aquifer flow modelling** Aquifer flow from areas of high pressure to areas of lower pressure is common. This will effectively increase the potential for dissolution of CO$_2$. This might be considered a benefit; however, if as is likely, flow is towards the shallower part of the aquifer, this could provide a route for effective leakage of CO$_2$ either by flow to the surface or to a position, where a phase change to a gas is likely, when the buoyancy is increased significantly and the potential for vertical migration through cap rocks is increased.

(iv) **Solubility in groundwater** In most water–rock systems, CO$_2$ is the most abundant gas, but rarely exceeds 2 mol% in solution. CO$_2$ solubility in water decreases sharply from 25°C to 250°C and with increasing salinity. This means that CO$_2$ in saline aquifers in sedimentary basins is less efficient and implies that low saline aquifers are better sites for CO$_2$ disposal.

(v) **Reaction with host rock** Formation damage may occur as a result of the removal of carbonate cements from the aquifer sandstone due to the increase of CO$_2$ in pore fluids. The stability of the ground may be affected, if minerals are dissolved or if pressures greater than geostatic are applied. Fracturing will increase permeability and porosity, but may also damage the integrity, if the sealing barrier exists.

(vi) **Groundwater pollution** The contamination of potable groundwater would occur if CO$_2$ escaped as a result of a seal failure to the host formation, a leak in the injection system, or if unidentified pathways such as faults or fractures existed.

SCREENING CRITERIA FOR STORAGE IN SALINE AQUIFERS

Not all sedimentary basins are suitable for CO$_2$ storage; some are too shallow and others are dominated by rocks with low permeability or poor confining characteristics. Basins suitable for CO_2 storage have characteristics such as thick accumulations of sediments, permeable rock formations saturated with saline water (saline formations), extensive covers of low porosity rocks (acting as seals), and structural simplicity. In general, storage sites should have (i) adequate capacity and injectivity, (ii) a satisfactory sealing cap rock and confirming unit, and (iii) a stable geological environment.

Realistic and quantitative information of the characteristics of the sub-surface is needed to assess the feasibility of sites. Geophysical surveys and borehole data analysis are carried out to define the sub-surface geological structures.

Safe and reliable containment of CO_2 in geological structures begins with the assessment of the location, structural features and characteristics of the target reservoir, its sequence stratigraphy, and depositional model. Reservoir characterization in the regional scale has reasonably straightforward data and interpretive requirements. Two-dimensional seismic coverage and well log data provide adequate basis for regional, structural and physical mapping, which is suitable for strategic purpose. For specific site characterization, however, three-dimensional seismic coverage augmented by downhole samples forms a minimum pre-requisite.

Saline aquifers are generally unused; therefore, documentation of the properties of the sub-surface is not compiled in an easy to access format. Today no standard methodology prescribes how a site for storage in saline aquifer must be characterized. However, site-specific data sets peculiar to a particular geological setting based on geological site descriptions from wellbores and outcrops are needed to characterize the storage formation and properties of the seal.

The following are the screening criteria proposed for assessing the feasibility of the storage site:

(i)   Safe and reliable containment of CO_2 begins with the assessment of the location, sequence, stratigraphy, structural features, and characteristics of the reservoir cap rocks.

(ii) The injection sites must be selected to have the geological properties that will assume that the CO_2 will remain trapped in the sub-surface isolated from the atmosphere for thousands of years.

(iii) The capacity of the storage site is the volume of pore spaces in the aquifer that could be occupied by CO_2. The reservoir must be large enough to be able to store the quantities of CO_2 planned.

(iv) Parameters must be sufficiently high to provide sufficient volume for the CO_2 and to allow injection of the CO_2. If the permeability of the rock is low or there are barriers to fluid flow, the rate at which CO_2 can be injected will get limited and may ultimately limit the amount of CO_2 that can be practically stored. Highly structural compartmentalized reservoirs are less suited to CO_2 storage than large unfaulted or high permeability reservoirs.

(v) The target formations should be greater than 800 m deep (bgl) as at this depth, the temperature and pressure conditions are such that the CO_2 exists in dense phase as a liquid and occupies much less pore volume than in gaseous state.

(vi) Potential storage sites should be in a geologically stable area as tectonic activity could create pathways for the CO_2 to migrate out of the reservoir through cap rock and the overlying strata and eventually to the surface.

(vii) It should have adequate separation from potable water.

Assuming that basement rocks would not have sufficient injectivity, thickness of sedimentary cover provides initial index for prospecting suitable formations. Younger sedimentary basins are more suitable as high porosity tends to be preserved at shallow depths. In older basins, porosity is lost due to cementation and compactness, because of depth of burial.

Several ongoing and planned projects for CO_2 storage in deep saline aquifers presently in progress are: Sleipner, Snohvit (Norway); Frio, Teapot Dome (USA); Gorgon, Otway (Australia); Minami Nagoaka (Japan); and Ketzin (Germany). These when completed will provide further insights about CO_2 storage in saline aquifers.

DISTRIBUTION OF SALINE AQUIFERS

World Scenario

Saline aquifers are quite widespread and present in almost all the onshore and offshore sedimentary basins occurring in different parts of the world as depicted in Figure 3.

From a capacity prospective, saline aquifers offer a significant potential and are suitable for any emission scenario. Storage capacity for different geological options is depicted in Table 1.

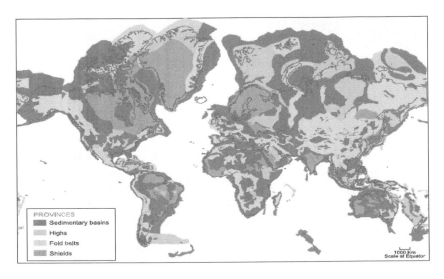

Figure 3 *Distribution of sedimentary basins around the world*

Table 1 : Global storage capacity for geological storage options

Geological storage option	Global capacity	
	Reservoir type lower estimate of storage capacity (Gt CO_2)	Upper estimate of storage capacity (Gt CO_2)
Depleted oil and gas fields	675	900
Unminable coal seams	3–15	200
Deep saline reservoirs	1000	Uncertain, but possibly 10^4

Indian Scenario

Almost two-third parts of the country occupied by the Archean igneous and metamorphic rocks have negligible porosity. In the Deccan basalts, only secondary porosity is preserved within the fractured zones and weathered layers. The porous formations are confined to semi-consolidated and consolidated sedimentary rocks occurring in the Ganga basin, Gondwanas, Cuddalore sandstone and their equivalents, in Rajasthan basin, Vindhyan basin, and so on (Figure 4).

The deep saline aquifers may prove to be an efficient option for carbon sequestration due to their wide distribution and proximity to the major emitting sources. The deep saline aquifers are present in different geological formations as revealed by exploratory drilling

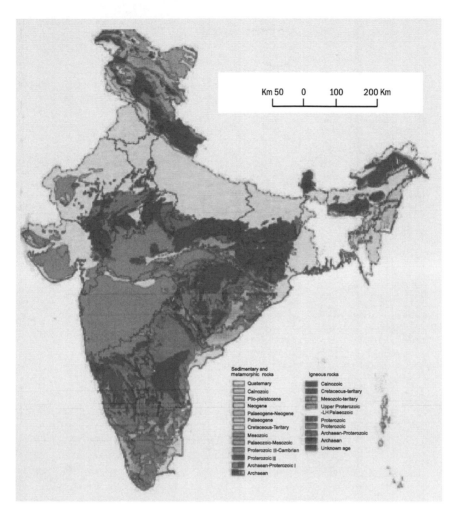

Figure 4 *Geological map of India*
Source: Geological Survey of India

for the delineation of deep aquifer zones. The distribution of inland salinity/saline aquifers in India is given in Figure 5.

Maximum saline areas fall in parts of Rajasthan, Haryana, Punjab, Uttar Pradesh, Gujarat, and Tamil Nadu. Incidentally, major thermal power plants are located in these areas.

In Gujarat, an area of 28,000 km^2 is affected by salinity. Exploration for deep saline aquifers was carried out up to 621 m bgl. The exploration indicated the presence of thick saline aquifers in the basalt and in Pre-Cambrian shale formations the aquifers are under free flowing conditions. In the Kutch region (coastal Gujarat), the saline aquifers were encountered at different depths and continue up to 458 m below

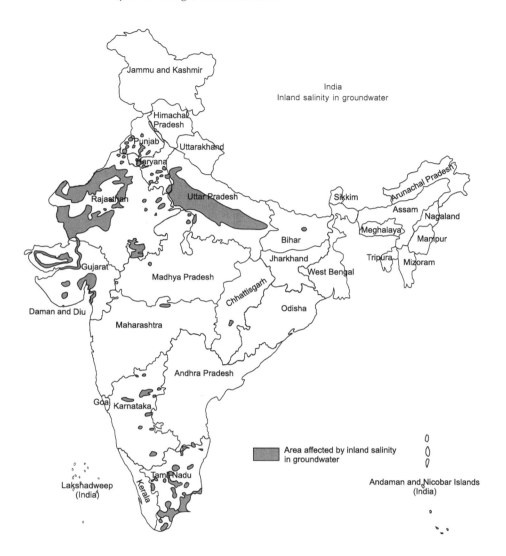

Figure 5 *Distribution of saline aquifers*
Source: Central Ground Water Board

ground level with the electrical conductivity (EC) value of 10,000 μ Siemens per cm.

In southern coastal areas of Tamil Nadu, the deep saline aquifers occur in the tertiary sandstone and alluvial formations over an area of 3750 km².

In the arid areas of Rajasthan, the exploration shows the presence of saline aquifers in about 1,000,000 km² and the salinity increases with depth. The exploration up to 610 m indicated the presence of granular zones with high salinity in Bikaner–Jaisalmer area.

In the Ganga basin, exploratory drilling and geophysical logging show the presence of deep saline aquifers. The saline aquifers are present in its western extension almost running for 342 km from Meerut to Rasalpur. The highly saline aquifers at all depths are present in an area of 8600 km^2 in Meerut and Agra districts. In the Meerut district, the saline aquifers have been traced up to 600 m bgl. The geophysical logs of 172 boreholes in the study area show the thickness of saline aquifers ranging from 30 m to 300 m indicating the presence of thick granular zones within the Ganga basin sediments.

PRESENT STUDIES

A tract of high salinity spread over an area of over 8600 km^2 occurs in western parts covering Ghaziabad, Faridabad, Agra, and Mathura districts. Based on the presence of salinity and sub-surface data from the exploratory boreholes, three areas in the vicinity of National Capital Region (NCR) in the Ganga basin were selected for deep resistivity surveys to comprehend the characteristics of the saline aquifers at depths of 800 m bgl and beyond (Figure 6).

The study indicated that the saline zones present within the alluvial sediments of Ganga basin in the study area are often discontinuous and thin, and occur at shallow depths, hence cannot be considered for sequestration of CO$_2$; occurrence of fracture zones saturated with saline water within the Alwar quartzites is recorded in the Pirthala-Tumsara area. These zones lack the depth, continuity, and characteristics of a

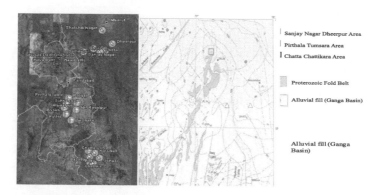

Figure 6 *Location map of the study area and regional geological setup*

suitable reservoir. The interpreted 60–110 m thick zone saturated with brackish water (EC values > 2000 ppm) within the upper Bhander sandstone of Vindhyan super group occurring at depth of 700–920 m bgl confined from Chattikara in the south to Chatta in the north can be a potential storage site, which warrants further studies.

As part of International Energy Agency (IEA) greenhouse gas programme, regional assessment of the potential geological CO_2 storage sites in Indian sub-continent has been carried out. The CO_2 storage potential of India's sedimentary basin in this report is far from definitive and the basins are classified as good, fair, and limited.

The Wuppertal Institute for Climate, Environment, and Energy in their final report on Prospects of Carbon Capture and Storage Technologies in Emerging Economies (Part II) Country Study: India, mentions that the storage capacity between 47 Gt and 572 Gt of CO_2 shows the strong uncertainty surrounding the studies presented. The calculation of an effective capacity would be even more speculative. In the first place, more knowledge about site-specific geology is needed. With such knowledge, the behaviour of CO_2 in the underground could be analysed to estimate storage efficiency. This will take years because a lot of work would have to be carried out.

CONCLUSION

Geological storage of CO_2 for reducing its emissions for mitigation of global warming is a new research area. There are gaps in our knowledge as to the regional storage capacity and potential of different sedimentary basins and the deep saline aquifers occurring within them. There is a need to prepare an inventory of the prospective sedimentary basins and database on the deep saline aquifers to assess their suitability for storage of CO_2.

In general, a storage site should have adequate capacity, a satisfactory sealing cap rock and confining unit, and stable geological environment. However, these are most difficult to assess for saline aquifer due to limited well-geophysical data. However, deep exploratory drilling and geophysical surveys carried out for oil and gas by Oil and Natural Gas Corporation Ltd (ONGC), and oil companies, water development projects by Central Ground Water Board, some sub-surface data about saline aquifers has been made available.

Further extensive research is needed both regionally and globally to study their true potential. In particular, the geology of these reservoirs needs to be researched in detail along with the integrity of their cap rock, connectivity to other aquifers, and the potential for potable water contamination.

At present, there appears to be no insurmountable technical barriers for geological storage of CO_2 as an effective mitigation option.

Summary

The scientific evidence is now overwhelming that the increased concentration of CO₂ in the atmosphere results primarily from the combustion of fossil fuels to meet energy needs of different sectors. Since, fossil fuels would be the mainstay for energy generation, CO₂ emissions are likely to increase exponentially in the coming decades. To contain and ultimately reduce its emission, it will be important to adopt cleaner technologies, capture, reuse, and store CO₂ in geological sinks.

In little over a decade, geological storage of CO₂ has grown from a concept to a potentially important mitigation option. At present, geological storage of CO₂ in saline aquifers seems to be a viable option. Saline aquifers occur almost in all sedimentary basins, are much more widespread, and could lie in the proximity of stationary emitting sources. In this chapter, the advantages and constraints of geological sequestration of CO₂ in saline aquifers are described. Saline aquifers are generally unused and of no economic significance, so there is no value addition from CO₂ storage. Studies have been undertaken to understand the rubrics of geological storage in saline aquifers to make a beginning. Here, the Indian perspectives are discussed.

BIBLIOGRAPHY

1. Bhandari, A. K. 2007. Study on identification of deep underground aquifers and their suitability for carbon dioxide sequestration. *Project Report DST (IS-STAC)*, 1–57

2. Bradshaw, J. 2006. Geological storage assessment examples from studies in Australia. *Technical Workshop on Carbon Dioxide Capture and Storage: India-Australia*, New Delhi

3. Bentham, M. and G. Kirby. 2005. CO_2-storage in saline aquifers, oil and gas science technology-Rev.IFP. *Dossier* 60(3):559–567

4. Central Ground Water Board. 2002. *Hydrogeology and Deep Groundwater Exploration in Ganga Basin in India*, CGWB, Faridabad, pp. 1–104. Central Ground Water Board. *Inland Groundwater Salinity*

in India, D. K. Chadha and S. P. Sinha (eds), CGWB, Faridabad, pp. 1–89

5. Department of Science & Technology. 2006. *Proceeding of the Second Inter Sectoral Meet on Technical Issue on Carbon Dioxide Sequestration,* Government of India.

6. Holloway, S., A. Garg, M. Kapshe, A. Deshpande, A. S. Pracha, S. R. Khan, M. A. Mahmood, T. N. Singh, K. L. Kirk, and J. Gale. 2008. A regional assessment of the potential for CO_2 storage in the Indian subcontinent, Commissioned Report. *British Geological Survey,* pp. 190

7. Goel, Malti, S. N. Charan, and A. K. Bhandari. 2008. CO_2 sequestration: Recent Indian research, in glimpses of geosciences research in India. *INSA Report* 55–60

8. Shahi, R. V. 2007. India's position on CCS technology in the context of climate change. *Inaugural Address of the International Workshop Carbon Capture and Storage,* NGRI, Hyderabad

9. Prospects of carbon capture and storage technologies (CCS) in emerging economies. 2012. *Report* of Wuppertal Institute, Germany (GIZ-PN 2009.9022.6) Deitsje Geselschaft fur Yechnishe Zusammenarbeit (GTZ) GmbH, Germany

10. Bradshaw, J. and T. Dance. 2005. Mapping geological storage prospectivity for the world's sedimentary basins and regional source to sink matching. In *Greenhouse Gas Control Technologies, 7th International Conference on Greenhouse Gas Control Technologies,* E. Rubin, D. Keith, and C. Gilboy (eds), 5–9 September 2004, Vancouver, Vol 1, Issue 4, pp. 583–592

11. U.S. Geological Survey. 2001. Mineral commodity summaries 2001. *U.S. Geolocial Survey,* p. 193

CHAPTER 8

Options for CO_2 Storage and its Role in Reducing CO_2 Emissions

B. Kumar

Ex-Chief Scientist, National Geophysical Research Institute,
Hyderabad-500007
Visiting Scientist, Gujarat Energy Research and Mangement Institute,
Gandhinagar-382007
E-mail: baleshk@yahoo.com

CARBON STORAGE IN GEOLOGICAL FORMATIONS

Carbon storage is a new worldwide initiative for capturing carbon dioxide (CO_2) from its point sources mitigating the impact of global climate change, which is one of the greatest environmental challenges the world community is facing. The capture of pure CO_2 from fossil fuel-based thermal power and industrial plants, which are the largest point sources of greenhouse gases (GHGs), and its storage in the geological formations such as depleted hydrocarbon reservoirs for enhanced oil and gas recovery; unmineable coal seams for coalbed methane (CBM) recovery; saline aquifers and basalt formations are the viable options (Figure 1). CO_2 is stored either in mineral, miscible and/or immiscible phase, and the storage security is high to medium and low. The salient features of different carbon storage mechanisms are discussed below.

Storage of CO_2 in Depleted Oil Reservoirs

The storage of CO_2 in the depleted oil reservoirs is an economically viable option, because the cost of CO_2 capture and transport is partly adjusted against the enhanced oil recovery (EOR). The CO_2 injected for the EOR can be stored in either miscible or immiscible phase and depends primarily on the pressure of the injection gas into the reservoir.

(i) **Miscible phase** In CO_2-EOR, the CO_2 mixes with the crude oil causing it to swell and reduce its viscosity, whilst also increasing or maintaining reservoir pressure. The combination of these processes

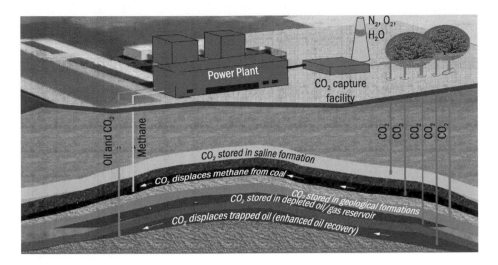

Figure 1 *Viable options for CO_2 storage*

enables more of the crude oil in the reservoir to flow freely to the production wells from which it can be recovered.

(ii) Immiscible phase In CO_2-EOR, the CO_2 is used to re-pressure the reservoir as a sweep gas to move the oil towards the production well.

The Weyburn-Midale CO_2 monitoring and storage project in Canada is one among the largest on-going projects for Carbon Capture and Storage (CCS) in the world. The Encana Cooperation has been injecting 5000 tonnes of CO_2 per day into the Weyburn oilfield for the dual purpose of EOR and for CO_2 storage. At the end of the first phase, the project conclusively demonstrated the suitability of CO_2 storage in the Weyburn field, while increasing the field's production by an additional 10,000 barrels per day. About 30 million tonnes (Mt) of CO_2 will be injected and permanently stored over the life of the project producing at least 130 million barrels of incrementally recovered oil.

In India, the Oil and Natural Gas Corporation Ltd (ONGC) has proposed CO_2-EOR for Ankleshwar oilfield in western India. The project has the potential to enhance the oil recovery as well as sequester 5 to 10 Mt of CO_2 during the project life.

Enhanced Coalbed Methane Recovery

Carbon dioxide sequestration in the unmineable coal seams serves the dual purpose that is, CO_2 storage as well as enhanced coalbed methane

(ECBM) recovery. Coalbeds typically contain large amounts of methane (CH$_4$) rich gas that is adsorbed onto the surface of the coal. The injected CO$_2$ efficiently displaces CH$_4$ as it has greater affinity to the coal than CH$_4$ in the proportion of 2:1, and is preferentially adsorbed displacing the CH$_4$ sorbed in the internal surface of coal layers. India has large potential for ECBM. The following factors affect desorption of CH$_4$ and the adsorption of CO$_2$:

(i) **Coal rank** The adsorption capacity is a function of the amount and reactivity of surface area contained in pores and the rank is known to have an effect on the amount of CO$_2$ that can be adsorbed into the coal porosity.

(ii) **Coal composition** Gases are adsorbed mostly by vitrinite rich facies of coal with low amount of minerals.

(iii) **Geologic age** Coal of different ages has different compaction that can affect pore size and distribution of pores thus affecting adsorption capacity.

(iv) **Depth/thickness** Greater depth/thickness provides greater volumes for coal seam storage.

Basalt Formations

The basalt formations are promising options for environmentally safe and irreversible long-time storage of CO$_2$. The CO$_2$ is injected in the supercritical state in which it has the physico-chemical properties between those of liquid and gas. Three main trapping mechanisms by which the CO$_2$ is stored in the basalts are the hydrodynamic, solubility, and mineral trappings. In hydrodynamic trappings the CO$_2$ is stored as a separate phase and has lower storage integrity. In solubility trapping, the CO$_2$ is dissolved in the formation fluids, providing a better trapping efficiency. The most effective is the mineral trappings in which the CO$_2$ reacts with the formation minerals and gets converted to geologically stable mineral carbonates.

The basalts are attractive storage options as they provide solid cap rocks and have favourable chemical compositions for the geochemical reactions to take place between the CO$_2$ and the formation minerals, rendering high level of storage security. The inter-trappeans between basalt flows also provide major porosity and permeability for considerable storage capacity.

Large igneous provinces like the Columbia River Basalt Group (CRBG) of the USA and the Deccan traps of India are the potential host media for the geologic storage of CO_2. The laboratory study of the *in-situ* mineralization on basalt shows fast mineralization reactions in terms of geological timescale (<1000 days). A feasibility study for the evaluation of basalt formations of India for environmentally safe and irreversible long-time storage of CO_2 has been carried out by National Geophysical Research Institute, Hyderabad in collaboration with Department of Science and Technology, New Delhi and Pacific Northwest National Laboratory, USA. Deccan traps with favourable geological, geophysical, and geochemical properties show promising options for carbon sequestration. The tectonically stable traps have intra-trappeans in between the flows for the enhancement of porosity and permeability.

However, Deccan basalts of India lie in the type II and III seismic zones of India and are overlain by thick Mesozoic sediments at places, which are hydrocarbon bearing. Therefore, storing CO_2 below Deccan basalts have to be dealt with lot of caution.

Saline Aquifers

Saline aquifers at depths of greater than 800 m also provide viable options for the storage of CO_2. Brine filled aquifers with high content of dissolved solids have little or no economic and social importance. The potential global storage capacity is also vast and is estimated to be nearly 1×10^{13} tonnes of CO_2. The high porosity and permeability of the aquifer sands along with low porosity cap rock such as shale provides suitable conditions for CO_2 storage. Over time, CO_2 gets dissolved in the brines and also reacts with the pore fluids/minerals to form geologically stable carbonates. The Sleipner project at Norway demonstrates the successful storage of CO_2 in the Utsira aquifer below the bed in the North Sea. About 1 Mt of CO_2 separated from natural gas, produced from Sleipner West, is being injected in the Utsira aquifer per year since 1996.

In 2006 the Department of Science and Technology in India initiated feasibility studies aiming at identification of deep underground saline aquifers and their suitability for CO_2 sequestration. Central Ground Water Board and Geological Survey of India have established the presence of saline aquifers up to depths of 300 m below ground level in the Ganga basin. Deep resistivity studies carried out at nine sites around New Delhi have also shown the presence of saline aquifers at depths of 800 m and beyond around Palwal and Tumsara.

INNOVATIVE CARBON STORAGE ADVANCES

Bio-Carbon Capture and Storage

Bio-carbon capture and storage (Bio-CCS) implies a close loop in which CO_2 is removed from the atmosphere by plants and these plants are converted into bio-fuels and return the part of that carbon to the atmosphere after the fuel is burned to generate energy. The use of bio-fuel production/fermentation process with CCS can provide an opportunity to reduce atmospheric CO_2 levels significantly. IEA-GHG (2011) study on the "Global Potential for Biomass and CCS" projected that the negative emissions potential realized by 2050 may be 3–4 giga tonnes (Gt) per year based on the present bio-fuel feedstock.

The European Joint Task Force on Bio-CCS has developed a new concept, known as Geo-Green (Figure 2). The main purpose and driving force for the work in the task force is to develop carbon negative options and find the system that withdraws more CO_2 from the atmosphere than it emits. Researchers are working to understand how to convert the new biomass into fuels, which could then exhibit the efficiency comparable to fossil fuels.

Bio-sequestration

In bio-sequestration, the pure CO_2 from power plant is sent to a photo bio-reactor (Figure 3) along with the nutrients to convert CO_2 into biomass and oxygen. The photo-reactor makes use of the natural process

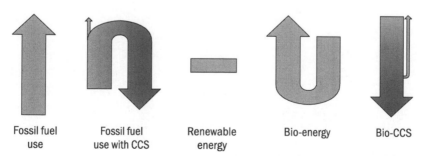

| | Fossil fuel
use | Fossil fuel
use with CCS | Renewable
energy | Bio-energy | Bio-CCS |

Figure 2 *Geo-Green, concept of European Task Force on Bio-CCS*
Source: www.geo-green.eu

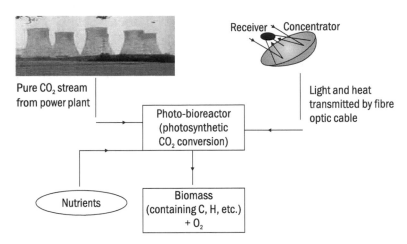

Figure 3   *Photosynthetic conversion of CO_2 into biomass*
Source: Energy Conversion and Management Vol. 46, February 2005, 403–420

'photosynthesis' to convert light, heat, and CO_2 to useful products such as carbohydrates and oxygen.

$$6\ CO_2\ (aq.) + 6\ H_2O\ (l) + light + heat \rightarrow C_6 H_{12} O_6\ (aq.) + 6\ O_2$$

The light and heat energy is supplied using solar energy parabolic concentrator/Fresnel lens devices and fibre optical light delivery system. The biological organisms like cyanobacteria or microalgae are used as nutrients. Considering carbon uptake of nearly 1.5 g/day in particular, a few thousand tonnes of CO_2 can be sequestered per year from the environment.

Geothermal Power from CO_2

The conventional geothermal energy is mostly generated by circulating potable water, but in the arid and semi-arid areas, where water is scarce, CO_2 can be used as geothermal heater fluid. CO_2 is cycled through a heat exchanger (Figure 4) in a 'Binary Geothermal System' as it extracts heat more efficiently than water. The heat exchanger creates steam for the turbines, while the CO_2 is compressed and re-injected into the geological formation in a close-loop system, stored and kept away from entering the atmosphere.

Increasing the Fertility of Soil and Ocean for CO_2 Uptake

As soil and ocean work as a sink for CO_2, researchers are working on the microbes that can increase the uptake of CO_2 by soil and ocean.

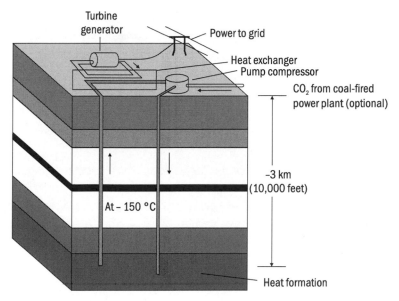

Figure 4 *Geothermal power using CO₂ as the heater fluid*
Source: www.ecogeek.org

These challenges once met will address the global climate issue most effectively.

The experts at Newcastle University, UK have discovered that sea urchins use nickel ions to capture CO_2 from the sea to grow their exoskeleton and this could be the key to convert atmospheric CO_2 into harmless solid carbonate minerals.

INDIAN INITIATIVES AND RECOMMENDATIONS

India stands with global community for accelerating research and development (R&D) in CCS technologies for sustainable energy future. However, CO_2 storage R&D in India is still at the initial stage and there is a need to develop programmes on CO_2-EOR, CO_2-ECBM, geologic storage in saline aquifers, and so on. India also has the need to initiate R&D on innovative carbon storage advances such as Bio-CCS, Bio-sequestration, and finding the mechanisms to increase the fertility of soil and ocean for CO_2 uptake.

Summary

The viable options for CO_2 storage are: CO_2-EOR; CO_2-EGR (enhanced gas recovery); geological CO_2 sequestration in basalt formations; and *saline aquifers. In India, Oil and Natural Gas Corporation Ltd has already initiated a project on CO_2-EOR in Ankleshwar oilfield of western India. The CO_2-EGR is still at R&D stage. Indian basalts may not be attractive proposition for carbon storage, as areas having basaltic rocks are prone to increased seismicity. Further, these rocks are underlain by thick Mesozoic sediments, which are light gaseous hydrocarbon bearing. Report on the occurrence of deep saline aquifers away from the coastal zones and R&D efforts in this area are scanty.*

The innovative carbon storage advances have been: Bio-CCS; and getting geothermal power with CO_2 instead of water in the arid areas, where water is scarce; and increasing the fertility of ocean and soil for CO_2 uptake. In this chapter, existing and innovative carbon options are discussed.

BIBLIOGRAPHY

1. Kumar, B., D. J. Patil, G. Kalpana, and C. Vishnu Vardhan. 2004. Geochemical prospecting of hydrocarbons in frontier basins of India. *Search and Discovery Article #10138 AAPG Online Journal*

2. Ministry of Power. 2007. *Carbon Capture and Storage Technology R&D, Initiatives in India.* New Delhi: DST

3. Cents, A. H. G., D. W. F. Brilman, and G. F. Versteeg. 2005. CO_2 absorption in carbonate/bicarbonate solutions: The Danckwerts criterion revisited. *Chemical Engineering Science* 60(21):5830–5835

4. Hongguan, Yu, Z. Guangzhu, F. Weitang, and Y. Jianping. 2006. Predicted CO_2 enhanced coalbed methane recovery and CO_2 sequestration in China. *Coal Geology,* doi: 10.1016/j.coal.10.002

5. Goel, Malti, B. Kumar, and S. Nirmal Charan (eds). 2008 *Carbon Capture and Storage R&D Technologies for Sustainable Energy Future,* p 222. New Delhi: Narosa Publishers

6. International Institute of Applied System Analysis, Austria. 2012. *Global Energy Assessment Report,* p. 118

7. International Energy Agency (IEA). 2006. *World Energy Outlook.* Paris: OECD/IEA

8. International Energy Agency (IEA). 2006. *Energy Technology Perspective: Scenarios & Strategies to 2050.* Paris: OECD/IEA

9. IPCC Special Report on Carbon Dioxide Capture and Storage. 2005. *Prepared by Working Group III of the Intergovernmental Panel on Climate Change.* Cambridge: Cambridge University Press. Details available at http://www.ipcc.ch/activity/srccs/index.htm

10. Kumar, B. 2007. Carbon dioxide capture & storage in emerging economics (Invited Talk), *G8 IEA/Carbon Sequestration Leadership Forum Assessment Workshop* 21–22 June, Oslo, Norway

11. Kumar, B. 2007. Carbon sequestration (Invited Lecture), *Indian Science Congress,* 3–7 Jan, Chidambaram, Tamil Nadu

12. Kumar, B. 2007. Carbon capture and storage R&D initiatives for sustainable energy future. *Proceedings of Indian Oil & Gas Review Symposium,* Mumbai, 3–4 September

13. MacGrail, B. Peter. 2006. Potential for carbon dioxide sequestration in flood basalts. *Journal of Geophysical Research* 111; doi:10.1029/2005JB004169

14. Tomski, P. 2006. Early commercial opportunities for carbon capture and storage (CCS) systems. *International Energy Agency,* Unpub. Rep., December

15. Kumar B., Nirmal Charan, and Malti Goel. 2007. International workshop on R&D challenges in carbon capture and storage technology for sustainable energy future. *Journal of the Geological Society of India* 70(1):174–175

CHAPTER 9

CO$_2$ Sequestration Potential of Indian Coalfields

Ajay Kumar Singh and Debadutta Mohanty

Methane Emission and Degasification Division,
CSIR-CIMFR, Dhanbad-826015, India
E-mail: ajay.cimfr@gmail.com

INTRODUCTION

Capturing carbon dioxide (CO$_2$) and storing it in unmineable coal seams in combination with enhanced coalbed methane (CBM) recovery provide an opportunity to stabilize the CO$_2$ concentration in the atmosphere and harness a clean source of energy. It is particularly important for countries like India, endowed with large coal deposits. CO$_2$ storage potential of coal seams is yet to be established. Friedmann *et al.* [1] suggested that injection of CO$_2$ within the coalbeds to replace methane (CH$_4$), and, thereby, enhance CBM recovery are likely to minimize CH$_4$ release in the environment, while CO$_2$ is stored.

Coal in India occurs in two stratigraphic horizons, namely Permian sediments mostly deposited in intra-cratonic Gondwana basins and early tertiary near-shore peri-cratonic basins and shelves in the northern and north-eastern hilly regions of Eocene-Miocene age. Gondwana basins of Peninsular India are disposed in four linear belts following several prominent lineaments in the Pre-Cambrian craton. In the extra-peninsular region (Darjeeling and Arunachal Pradesh), Lower Gondwana sediments occur as thrust sheets overriding Siwalik sediments. Besides coal, lignite deposits in younger formations also occur in the western and southern parts of India.

Assessment of CO$_2$ sequestration potential based on CO$_2$ and CH$_4$ sorption capacities of coalbeds in Indian sedimentary basins, properties of potential coalfields, and suggestions for future work are discussed in this chapter.

COAL AND LIGNITE DEPOSITS IN INDIA

Coal formation in India is continuation of Great Gondwana formation extending over Antarctica, Australia, South Africa, and South America (called Gondwana). Super group of the Indian Peninsula comprises 6–7 km thick clastic sequence formed in Paleozoic to Mesozoic era during Permian to Cretaceous period. Originating from highlands of central India, the Gondwana sedimentation occurred in graben or half-graben trough alignment due west to east and two parallel drainage channels north–west to south–east. Singrauli basin of Son valley occupies the junction between the east–west trending Damodar-Koel graben and the northeast–southwest trending rift zone of Son Mahanadi valley. Son valley lies to the immediate west of Damodar and north-west of Mahanadi valleys. The Wardha, Godavari, and Satpura Gondwana valleys lie in the southern and central part of India, respectively. A brief account of the pattern of sedimentation in Gondwana basins has been presented by Casshyap [2].

Gondwana sedimentation, the main repository of coal in India, is divided in Lower Gondwana corresponding to Lower and Upper Permian and Upper Gondwana corresponding to Lower Cretaceous and Lower Jurassic age. Acharyya [3] has reported that the Lower Gondwana belts are controlled by Pre-Cambrian crustal structure such as Archeancratonic sutures and Protozoic mobile belts. The formation lineament followed east–west trending zone occupied by Damodar, Koel, Son, and Narmada rivers. Northwest–southeast trending zone is occupied by Mahanadi river in the north and Godavari and Wardha rivers in the south. Another north–south trending zone north of Damodar river delineates Rajmahal group of coalfields that flank the Rajmahal hills. The coal and lignite deposits in Indian sedimentary basins are exhibited in Figure 1.

Nearly 99.7% sub-bituminous to bituminous coal of India is available in the Lower Gondwana in the eastern region of India located in West Bengal, Jharkhand and in Madhya Pradesh, Chhattisgarh, Odisha, Andhra Pradesh, and Maharashtra. There are more than 65 known basins in Lower Gondwana sediments spread over nearly 64,000 km^2. Excluding unproductive parts and small detached outliers, the potential coal bearing area is about 14,500 km^2 [4].

Tertiary formations are spread over the periphery of peninsula along the coast in Tamil Nadu, Kerala, Gujarat, and Himalayan foothills from Pir Panjal of Jammu and Kashmir to Abor Hills and Kuen Bhum range

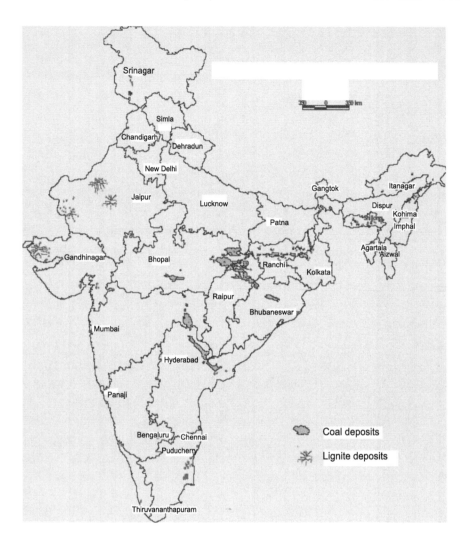

Figure 1 *Coal and lignite distribution in India*

of Arunanchal Pradesh. Deep-seated occurrence of lignite indicating substantial reserve of lignite at depth in the region of Kalol of Cambay basin, Barmer and Sanchor basins seems to be exceptional in nature.

COAL INVENTORY OF INDIA

Coal inventory of India is based on data made available from Geological Survey of India, Central Mine Planning and Design Institute, Mineral

Exploration Corporation Ltd, Singareni Collieries Company Ltd, and some other agencies. The coal seams over 0.5 m thickness and down to the depth of 1200 m are included for resource evaluation. Depth-wise reserve position in different states as available on 1 April 2012 [5] is given in Table 1.

EXTRANEOUS COAL DEPOSITS OF INDIA

Large tract of land in India is still not covered for regional exploration even in well-delineated basins. Many a times, surface exposures are misleading, terrain is unfriendly or exploration is technically difficult, and, hence, not included in potential coal basins. They are often identified by chance or in the process of exploration for oil and natural gases and

Table 1 : Depth-wise Gondwana coal resources of different states of India as on 1 April 2012

State	Resource estimate under depth			Total reserve (million Tonnes)
	0–300 m	300–600 m	600–1200 m	
Andhra Pradesh	9,654.17	8,754.74	3,745.95	22,154.86
Bihar	160.00	–	–	160.00
Chhattisgarh	37,755.15	11,609.08	1,481.92	50,846.15
Jharkhand	41,118.05	16,321.79	8,703.94	80,356.20**
**Jharia	-------14,212.42---------------			
Maharashtra	7,522.32	3,165.28	194.49	1,0882.09
Madhya Pradesh	15,367.65	8,656.20	352.41	24,376.26
North-eastern states	1,394.66	202.00	–	1,596.66
Odisha	46,653.67	23,086.73	1,707.01	71,447.41
Uttar Pradesh	1,061.80	–	–	1,061.80
West Bengal	13,155.05	12,127.36	5,333.31	30,615.72
Total	173,842.52	83,923.18	21,519.03	293,497.15
**Jharia	-------14,212.42---------------			
% share	59.23	28.60	7.33	100
**Jharia	-------4.84%------------------			

**Jharia coalfield reserve position is available in group of 0–600 m depth range only. The reserve position within 300 m depth over is found in all the active coalfields except a few hidden pockets under basaltic flow or soil cover
Source: Compiled by authors.

CBM. The coal reserve of such areas is not covered in National Coal Inventory by the Geological Survey of India.

The areas not covered under regional exploration for resource estimation could be the grey area thrown deep by way of faulting or concealed deposits under deep cover, loose alluvial, or igneous traps. Some of these areas have been included in the present exercise, because of their potential usage for CO$_2$ sequestration in the days to come.

GREY AREAS OF COALFIELDS

Well-delineated coalfields of nearly 37,000 km^2 area have a few patches under deep cover beyond the resource estimation and eventual mining limits. Such blocks occur under the influence of intra-basinal faults or subsidence of the basal formation. In some of the patches, the complete Damuda series are found, when the main coal repository – the Barakar formation occurs below 1500 m, while in other patches, even the top coal bearing formation, Raniganj is overlain by Upper Gondwana formations. The coals of such patches are not included in resource estimation. The details of these blocks are of interest, because of location, likely characteristics, their availability for CBM exploitation, and may be for CO$_2$ sequestration.

Jharia Coalfield

Jharia coalfield, the most extensively explored coalfield with nearly 95% proved reserve within 300 m depth cover, is explored up to 1200 m depth cover, with prime coking coal availability within 1050 m depth cover only [6]. Development of Moonidih mine just within the fringe area of Barren Measure to 510 m depth cover is intersecting XV seam, while the coking coal seam IX may be intersected in the shaft line at 830 m depth cover.

Barren Measure up to nearly full thickness of 730 m may have to be crossed to touch the coking coal seams along the periphery of Mahuda block, while within the Mahuda block overlain by Raniganj formation, the depth of the top-most Barakar seam may be 1800–2000 m deep beyond mining limit as per any imagination.

Raniganj Coalfield

As the fully developed Raniganj formation has 1035 m thickness in the Raniganj coalfields, hardly any effort has been made to touch Barakar formation underneath 350 m thick Barren Measure along the northern periphery of Raniganj formation exposure.

Sohagpur Coalfield

Sohagpur coalfield is divided into northern and southern sections by Chilpa Bahmani Fault of maximum 600 m throw trending east to west running from Burhar to Sonhat coalfields. As a result, exposed Barakar formation in the southern part is thrown down in the northern portion of the coalfield. The portion is further complicated with the exposure of Supra Barakar, Lameta, and Deccan trap flow of Upper Cretaceous age in the northwestern part [7]. Raniganj formation of 525 m thickness has a number of thin coal seams and dolerite sill of variable thickness [8].

The Barakar formation in this section is buried below 1200–2000 m and its geological details are not available. Middle Barakar of 150–250 m, the main repository, contains all the important coal seams. The cumulative thickness of the seams is 15–20 m, while in Upper Barakar, the seams in the eastern part has gained 60 m thickness at places.

South Karanpura Coalfield

South Karanpura coalfield in narrow elongated form is developed along the south boundary fault in folded basin with Bundi Basaria metamorphic inliers. The coalfield is a half-graben trough in semi-elliptical shape, traversed by basin margin fault, and a number of intra-basinal faults. Central part of the basin has all the sediments of Damuda group including 600 m thick Raniganj and nearly 300 m thick Barren Measure over the Barakar formation.

East Bokaro Coalfield

Bokaro coalfield, extending over 64 km in the east–west direction and maximum width of 11 km in the north–south direction, is divided in two parts by Lugu Hills in east and Bokaro coalfields in the west. The coalfield is traversed by a number of strike and oblique faults including prominent Govindpur Pichri Fault.

Talcher Coalfield

The Talcher coalfield, the southern-most member of the northwest–southeast trending Mahanadi valley, is the most important coalfield of Odisha. It falls mostly in the Dhenkenal district with a small part in the Sambalpur district.

CONCEALED COALFIELDS

Concealed areas are of recent discovery and invariably under deep cover, underneath basalt or alluvial cover. The exact extension of most of these basins is just indicative by way of geophysical prospective and with skeleton boreholes drilled for oil and natural gas exploration. The reserve estimates need correction with every new drilling record. The search for concealed coalfields is made with respect to Damuda sequence of formations within Gondwana sediments. It is, however, not certain that every patch with Raniganj formation will definitely have Barakar formation in sequence after crossing Barren Measure.

In most of the cases, the deposits follow established lineament and appear to be extension of active coal mining areas. The presence of South Gondwana formation in Bangladesh and intersection of thick coal seams at Singur under 1600 m depth cover support the extension of Damodar valley lineament beyond the present Raniganj coalfield. Concealed coalfields in major Gondwana lineament in Wardha and Godavari valley follow identical trend. This has been supported by geophysical logging and deep-hole drilling for oil exploration. The details i.e. area, seam distribution, and reserve estimate of coal in the case of deposits concealed under basaltic bed or alluvium are not available, but the process of their exploration by non-mining agencies in the near future is not ruled out.

UNMINEABLE COALBEDS IN WELL-DELINEATED COALFIELDS

The mining limit is decided with due consideration to quality, fuel value, market demand, market price, basin location, and abundance of coal. The limit as such varies for different grades of coal and location of the coalfields. There is no decided guideline for making futuristic extrapolation for mining limit, but in the light of past experience and future projection of global technological input, the following conservative suggestions are made. A buffer zone of 100 m cap rock is suggested between the mineable seams and unmineable coalbeds to contain gas injected in the coalbeds.

The mining in any basin will be marginally affected by distribution of different grades of coal and accordingly the operational limit will shift. For example, in the case of Jharia coalfield, the coking nature of the bottom-most zero seam may attract mining beyond the non-coking coal seams. I–V Superior grade coal in lower seams of Raniganj has always encouraged going deeper leaving aside the upper coal seams. Similarly, the operational limit of power grade coal accepted to be mined by opencast may alter in the case of coalescing of seam.

The storage characteristic of different beds, the possibility of CO_2 injection along with CH_4 drainage, and potential of gas sequestration should be estimated for unmineable coal seams.

COMPARATIVE ADSORPTION OF CO_2 AND CH_4

The coal mass has CH_4 in micropores invariably less than its adsorption capacity. Injection of CO_2 in such coalbeds has the option of occupying the void or occupies the total surface area by even displacing the CH_4 molecules. The studies conducted so far support the latter option, because of stronger affinity of CO_2 to the coal molecule. It has been found that with the displacement of each CH_4 molecule, 2 to 3 molecules of CO_2 are accommodated and, thus, its adsorption reaches closer to near completion.

The adsorption, storage, and generation of CH_4 are also known to depend upon the surface area of microporous system, thickness of coal seams, and the confining pressure. Methane sorption capacity for Indian coals has been investigated [8]. Based on the research of the last two decades, it has been generally accepted that coal can adsorb more CO_2 than CH_4 and that CO_2 is preferentially adsorbed onto the coal structure over CH_4 in the 2:1 ratio [9].

IDENTIFICATION OF POTENTIAL COALBEDS FOR CO_2 STORAGE

The potential sites for CO_2 storage in coalbeds of Indian territory have been identified with due consideration of accepted exploration norms, depth-wise resource distribution, quality-wise abundance, and mining status of coal. The following areas appeared to be promising sites for CO_2 storage (Table 2).

Properties of Potential Coalbeds

In the absence of knowledge of CH_4 adsorption capacity with CO_2 injection, empirical equations may be used for estimation of gas quantity in a particular coalbed. Coal characterization properties such as vitrinite reflectance percentage and proximate analysis are found to be the

Table 2 : Identified candidates for CO_2 storage in India

Category of coalbeds	Grade of coal	Candidates/basins
Unmineable coalbeds in well-delineated coalfields	Power grade coal	Singrauli, Mand Raigarh, Talcher, Godavari
Grey areas	Coking coal	Jharia, East Bokaro, Sohagpur, S. Karanpura
	Superior non-coking coal	Raniganj, South Karanpura
	Power grade coal	Talcher
Concealed coalfields	Tertiary age coal	Cambay basin, Barmer Sanchor basin
	Power grade coal	West Bengal Gangetic plain, Birbhum, Domra Panagarh, Wardha valley extension, Kamptee basin extension

Table 3 : Proximate analysis and rank of unmineable and grey area coalbeds

Coalfields	Basic parameters (mmf* basis)		Vitrinite reflectance
	VM (%)	FC (%)	Av. Ro (%)
East Bokaro	28–36	85–90	0.85–1.05
South Karanpura	37–40	80–84	0.60–0.80
Jharia-Barakar	17–35	87–93	0.90–1.30
Raniganj	39–44	79–90	0.70–0.85
Rajmahal-Barakar	38–40	78–81	0.45–0.50
Singrauli-Barakar	37–45	78–81	0.45–0.50
Sohagpur	34–40	80–87	0.55–0.65
Pench valley	32–40	82–89	0.50–0.60
Wardha valley	35–40	78–82	0.55–0.60
Godavari valley	35–42	78–83	0.55–0.60
Talcher	35–45	79–82	0.50–0.55

*mmf (mineral matter free basis)

relevant parameters affecting the gas recovery and subsequently CO_2 storage in coalbeds. These properties for the coalfields targeted for the said exercise are summarized in Tables 3 and 4. Some of the properties in the same coalfields varied in non-coking to coking coals and the average figure has been taken for approximate estimate.

Table 4 : Proximate analysis and rank of concealed coalbeds

Coalfields	Basic parameters (mmf* basis)		Vitrinite reflectance
	VM (%)	FC (%)	Av. Ro (%)
Cambay basin	45–58	52–68	0.32–0.44
Barmer Sanchor basin	47–60	48–66	0.26–0.40
Birbhum	16–38	68–86	1.10–1.86
Wardha valley	24–35	72–88	0.54–0.68
Kamptee Kanhan valley	26–36	75–92	0.52–0.66

*mmf (mineral matter free basis)

Storage Capacity of Potential Coalbeds

Carbon dioxide storage capacity of unmineable coal seams for different categories of coalfields has been discussed as follows:

- Unmineable coalbeds in well-delineated coalfields: In-depth coal resource analysis of Indian territory as per quality, depth-wise distribution, and status of exploration has supported in identification of suitable sites for CO_2 sequestration. The resources reported by Geological Survey of India and other agencies have been classed as mineable and unmineable on the basis of the following factors:
  (i) Exploration limit of coal has been to 1200 m depth cover.
  (ii) Coking and superior grade non-coking up to the explored limit has been classed as mineable.
  (iii) Inferior grade non-coking coal (Grade E–G) below 600 m depth cover.
  (v) Damodar and Mahanadi valleys have been taken as within mineable limit.
  (vi) Mineable limit for inferior grade non-coking coal of Godavari and Wardha valleys have been taken as 800 m due to premium pricing structure.

  The coalbeds of Singrauli, Mand Raigarh, Talcher, and Godavari valley come under this class, where the coal reserve is available below the mining limit. With a view to capping injected CO_2 in the coalbeds, minimum 100 m thick top formation is proposed to be left between the working horizon and the non-mining zone. The vitrinite percentage of these sites is low in the range of 40%–60%, vitrinite reflectance (Av. Ro%) within 0.4%–0.6% and ash within 15%–45%, average 35%. The seams according to these properties are sub-bituminous in rank with poor cleat frequency and aperture.

Table 5 : Unmineable coal reserve and CO$_2$ storage capacity

Coalfield	Estimated adsorption capacity of CO$_2$(m^3/t)	Coal reserve (Mt)	CO$_2$ storage capacity (Bm^3)	CO$_2$ storage capacity (Mt)	CO$_2$ storage capacity (90%) (Mt)
Singrauli	Average 20.0	37.0	0.74	1.46	1.32
Mand Raigarh	Range 16.0–23.0 Average 19.0	79.0	1.50	2.97	2.67
Talcher	Range 17.2–24.8 Average 20.4	1,017.0	20.80	41.18	37.06
Godavari	Range 16.8–22.2 Average 19.2	1,976.0	38.02	75.28	67.75

The coal reserve, CH$_4$ reserve, and CO$_2$ storage capacity for these sites are summarized in Table 5.

- Grey area coalbeds: The extension of coalbeds below 1200 m depth cover in coking and superior grade non-coking coal has not been explored even though the continuity of the coalbeds was well-indicated within the lineament. The coalbeds of such zones beyond mineable limit have been classed as Grey Area reserve. These reserves in the case of East Bokaro, South Karanpura, Jharia and Raniganj, and Sohagpur are below 1200 m depth cover, while in the case of inferior grade non-coking the limit is 600 m for Son Mahanadi valley and 800 m for Wardha Godavari valley coalfields. The coal and CBM, recoverable CBM, and CO$_2$ storage capacity for these areas are summarized in Table 6.

  The CH$_4$ reserve in these locations is within 3.15–11 Bm^3 in 76–450 km^2 area. Cumulative seam thickness is very high within 15–120 m and average gas content within 2.4–7.6 m^3/t of coal. Some of the seams of Damodar valley coal basins have gas concentration above 19 m^3/t of coal. Total CO$_2$ sequestration even with 60% CH$_4$ recovery is estimated over 114 BCM or 226 Mt approximately.

- Concealed coalbeds: These coalbeds left in resource estimation exercise, because of the basalt trap or thick alluvium beds, are classed as concealed coalbeds. The bottom-most coal bearing Barakar formation in such operations has been located within 300 m to 3 km depth cover over Nagaland to Cambay basin of Gujarat. For the CO$_2$ sequestration or even ECBM recovery, the beds below 2000 m

Table 6 : Estimated CO_2 adsorption capacity in grey area coalbeds

Coalfield	Estimated CO_2 adsorption capacity (m³/t)	Cumulative coal seam thickness (m)	Block area (km²)	Coal reserve (Bt)	CO_2 storage capacity (Bm³)	CO_2 storage capacity (Mt)	CO_2 storage capacity (90%) (Mt)
South Karanpura	Range 19.5–28.0 Average 24.5	73.0	76.0	0.75	18.35	36.33	21.80
East Bokaro	Range 22.3–33.5 Average 28.1	100.0	113.0	1.53	42.90	84.94	76.45
Jharia	Range 22.0–56.0 Average 34.5	40.0	193.0	1.04	35.96	71.20	64.08
Raniganj	Range 20.8–29.0 Average 24.0	30.0	240.0	0.97	23.33	46.19	41.57
Sohagpur	Range 18.9–26.4 Average 22.6	15.0	450.0	0.91	20.59	40.76	36.69
Talcher	Range 17.2–24.8 Average 20.4	120.0	149.0	2.41	49.24	97.49	87.75

Table 7 : Concealed area coal reserve and CO$_2$ storage capacity

Coalfield	Estimated adsorption capacity (m^3/t)	Cumulative thickness of the coal seams (m)	Area of the block (km^2)	Coal reserve (Bt)	CO$_2$ storage capacity (Bm^3)	CO$_2$ storage capacity (Mt)	CO$_2$ storage capacity (90%) (Mt)
Cambay basin	Range 13.8–19.6 Average 16.7	102.0	6,900	63.0	1,057.81	2,094.45	1,885.02
Barmer Sanchor basin	Range 128–18.4 Average 15.6	100.0	6,700	60.0	936.00	1,853.28	1,667.95
West Bengal Gangetic plain	Range 16.4–23.2 Average 18.3	–	–	7.2	131.76	260.88	234.80
Birbhum coalfield	Range 17.2–24.8 Average 20.2	100.0	312.0	4.2	85.08	168.46	151.61
Domra Panagarh	Range 18.6–25.8 Average 21.8	48.0	116.0	0.751	16.39	32.45	29.20
Wardha valley extension	Range 15.7–22.8 Average 17.8	13.0	212.0	0.37	6.62	13.11	11.80
Kamptee extension	Range 7.2–9.2 Average 8.1	14.0	300	0.57	9.81	19.42	17.48

have not been included in concealed potential sites. In case such sites are indicated roughly and the boundary and lithology are not defined, then they are also excluded from the present exercise for the time being. The representative gas content, coal rank, and CO_2 storage potential for these fields are based on information available for the nearest coalbed of the lineament or from different sources (Table 7).

The gas resources of the above coalfields as estimated are based on the presumption that the saturation level of the coal mass will be nearly 90% during the lifetime of the bore wells, with recovery of CH_4 as per best practice. The storage capacity of some of the candidates are insignificant, particularly those of Wardha Kamptee extensions and unless the limit is precisely delineated, may not be of any use. Similarly, the storage potentials of unmineable beds of Mand Raigarh and Singrauli are also insignificant and even if ignored, may not materially change the situation. Delineation of concealed coal basins not yet well defined may make difference in CO_2 storage capacity in future. The Barmer Sanchor basin finding is a clear example, the latest finding of which has improved the CO_2 storage potential.

CONCLUSION

Deeper unmineable coals are prospective candidates for CH_4 recovery with simultaneous sequestration of CO_2 in coalbeds thereby reducing the atmospheric greenhouse gases (GHGs) level and providing a clean source of fuel. Indian coalbeds may be classified into grey, concealed, and unmineable based on its depth of occurrence and grade characteristics. CO_2 sequestration potential in Indian coalbeds is estimated to be 4459 Mt. However, further research on selection of suitable candidates for CO_2 storage at a specific site demands for a detailed economic appraisal taking into consideration the daily CO_2 generation from the point sources and the total gas likely to be generated during the lifetime of the power station with the present rate of consumption of coal.

Summary

India's coal reserves include the Gondwana and tertiary coals and lignite deposits. These can serve as storage sites for CO_2 limited by permeability and other controlling parameters such as sorption capacity and porosity. In this chapter, the information on coal availability, storage capacity, and coal characteristics are used from the best published data sets for *assessment of CO_2 sequestration potential of unmineable coalbeds in India. The scientific assessment of CO_2 storage potential in coal bearing sedimentary basins may serve as a basis for research, planning and policy to enable carbon capture, and storage technology deployment in Indian coalfields.*

ACKNOWLEDGEMENT

The authors are grateful to Dr T. N. Singh, Former Director, CMRI, Dhanbad for his valuable advice in preparing this paper. Thanks are also due to Dr A. Sinha, Director, CSIR-CIMFR, Dhanbad for granting permission to publish this paper.

REFERENCES

1. Friedmann, S. J., J. J. Dooley, H. Held, and O. Edenhofer. 2006. The low cost of geological assessment for underground CO_2 storage: policy and economic implications. *Energy Conversion Management* 1894:47

2. Casshyap, S. M. 1979. Patterns of sedimentation in Gondwana basins. *Fourth International Gondwana Symposium,* Geological Survey of India, Calcutta, India, January 1977

3. Acharyya, S. K. 2000. Comment on "Crustal structure based on gravity-magnetic modelling constrained from seismic studies under Lambert Rift, Antarctica and Godavari and Mahanadi rifts, India and their interrelationship" by D.C. Mishra *et al. Earth and Planetary Science Letters* 595:179

4. Coal India Limited. 1993. *Coal Atlas of India.* Ranchi: CMPDI

5. Geological Survey of India. 2013. Inventory of coal and lignite resources of India as on 01.04.2012. Details available at http://www.portal.gsi.gov.in/gsiDoc/pub/coal_resource.zip

6. Chandra, D. 1992. *Jharia Coalfield.* Bangalore: Geological Society of India

7. Chandra, D., R. M. Singh, and M. P. Singh. 2000. *Textbook of Coal.* Varanasi: Tara Book Agency

8. Singh, A. K. 1999. Opportunities for coalbed methane exploitation in India. *Proceedings of International Conference on Mining Challenges of the 21st Century* November, New Delhi, The Institution of Engineers (India), p. 369

9. Chikatamarla, L. and M. R. Bustin. 2003. Sequestration potential of acid gases in western Canadian coals. *Proceedings of the 2003 International Coalbed Methane Symposium,* 5–8 May, University of Alabama, Tuscaloosa, Alabama, p.1

CHAPTER 10
Carbon Sequestration Potential of the Forests of North-eastern India

P. S. Yadava and A. Thokchom

Centre of Advanced Study in Life Sciences, Manipur University,
Imphal-795003, Manipur
E-mail: pratap_yadava@ymail.com

INTRODUCTION

Forests play a significant role in capturing carbon dioxide (CO_2) from the atmosphere through photosynthesis, converting it into forest biomass, and then releasing it into the atmosphere through plant respiration and decomposition. Therefore, forests contribute positively to global carbon balance. Carbon sequestration in forest soil and vegetation has been used for achieving greenhouse gas (GHG) reduction target. Kyoto Protocol Article 3.3 refers to net changes in carbon stock that occurred during 2008–12 as a result of afforestation, restoration and deforestation, which have taken place since 1990. Thus, forests play an important role in climate mitigation and adaptation. There is also a need for forest-dependent population and forest ecosystems to adapt to this challenge. Forests maintain high carbon stock by reducing deforestation and by promoting sustainable management of all types of forests. Sustainable forest management provides an effective framework for forest-based climate mitigation and adaptation.

As the forests store more CO_2 than the entire atmosphere, their role is very critical [1]. Carbon sequestration is an important part of the overall carbon management strategy to help in reduction and to mitigate global CO_2 emission. Therefore, it is essential to continue to emphasize the understanding of carbon cycle, which drives future global change scenario and impacts the efficacy of sequestration options in natural ecosystems.

The Kyoto Protocol under United Nations Framework Convention on Climate Change (UNFCCC) has international agreement on incorporation of forestry activities to this major environmental challenge. At the international level, Intergovernmental Panel for Climate Change (IPCC), Global Carbon Project of International Global Biosphere Programme and World Research Programme and at the national level, Vegetation Carbon Pool Programme of Department of Space, Government of India, National Action Plan on Climate Change and National Programme on Carbon Sequestration Research of Department of Science and Technology, New Delhi have emphasized understanding the role of forests in carbon capture and storage under anthropogenic change.

Among the world's forests, tropical forests have the greatest potential to sequester carbon primarily through reforestation, agroforestry, and conservation of existing forests [2]. Plantation may act as corridors, source or barrier for different species, and a tool for landscape restoration. The total carbon stock in Indian forests has increased over a period of 20 years (1976–2005) and amounts to 10.01 GtC [3]. Carbon stock in both soil and vegetation in the existing forests is estimated to be 8.79 GtC [4]. The forest ecosystems have long been subjected to many human-induced pressures such as land-use changes, over harvesting, fire, and so on. Climate changes constitute an additional pressure that could alter these ecosystems. Impact of climate change can affect the mitigation potential of forests either increasing (nitrogen deposition and CO_2 fertilization) or decreasing (negative impact of air pollution) the carbon sequestration [5].

The forests in north-eastern India act as a sink of carbon. When there is any increase in carbon stock, it is retained in the forest vegetation itself as well as in the form of organic matter in the soil. However, burning of forests for shifting cultivation and deforestation release CO_2 in the atmosphere. Therefore, vast areas of abandoned shifting cultivation in fact offer a great potential to carbon sequestration through restoration programme of reforestation and agroforestry in the region.

A number of studies have been reported on carbon stock in India on the basis of growing stock volume data of forest inventories using conversion factors [6–10]. However, limited information is available on carbon sequestration potential of the forests of north-eastern India [11–13]. Soil organic carbon stock in the north-eastern India has been investigated recently by Bhattacharya *et al.* [14] and Choudhury *et al.* [15]. The carbon sequestration potential of forests in terms of soil organic carbon, carbon stock, and rate of carbon sequestration in the aboveground biomass and soil CO_2 flux in different forest ecosystems of north-eastern regions has been discussed and its role in the reduction of CO_2 to the atmosphere.

STATUS OF FOREST COVER IN NORTH-EASTERN INDIA

The total forest cover of north-eastern India, comprising eight states, namely Arunachal Pradesh, Assam, Manipur, Meghalaya, Mizoram, Nagaland, Sikkim and Tripura, is 173,219 km^2, which is 66.07% of its geographical area as against the national average of 21.05% [16] (Table 1). The north-eastern region comprises 7.98% of the geographical area of the country, but accounts for nearly 25% of the forest of the country.

Out of the total forest cover, 14.73% is very dense, 44.29% is moderately dense, and 40.98% is open forests, respectively. However, during 2007–09, there was a loss of 549 km^2 forest cover. Manipur and Nagaland have lost the highest forest cover of 190 km^2 and 146 km^2, respectively. Almost all the states have recorded a decline in the forest cover except Sikkim, which has maintained its forest cover as such [16]. Forest covers in the tribal districts of the country have also declined to 679 km^2 during 2007–09 (Table 2). Shifting cultivation and dependence of tribal people on forests for their livelihood are the main factors in the decline of forest cover in the hilly region. The tribes have lived in forests in harmony since ages and some of the tribal districts have been identified under Integrated Tribal Development Schemes of the government.

Table 1 : Forest cover in the north-eastern states in 2009 [16]

(Area in km^2)

State/UT	Geographical area (GA)	Forest cover				% of GA	Change*	Scrub
		Very dense forest	Moderate dense forest	Open forest	Total			
Arunachal Pradesh	83,743	20,868	31,519	15,023	67,410	80.50	– 74	122
Assam	78,438	1,444	11,404	14,825	27,673	35.28	– 19	182
Manipur	22,327	730	6,151	10,209	17,090	76.54	– 190	1
Meghalaya	22,429	433	9,775	7,067	17,275	77.02	– 46	485
Mizoram	21,081	134	6,086	12,897	19,117	90.68	– 66	1
Nagaland	16,579	1,293	4,931	7,094	13,318	80.33	– 146	3
Sikkim	7,096	500	2,161	698	3,359	47.34	0	363
Tripura	10,486	109	4,686	3,182	7,977	76.04	– 8	72
Grand total	262,179	25,511	76,713	70,995	173,219	66.07	– 549	1,229

*In forest cover

Table 2 : Forest cover in tribal districts of India in 2007 and 2009 [16]

			Forest cover				% of GA	Change*	Scrub
State/ UT	No. of tribal districts	Geogra- phical area (GA)	Very dense forest	Mode- rate dense forest	Open forest	Total			(Area in km²)
2007									
Grand total	188	1,105,744	59,791	193,813	159,121	412,625	37.32	690	12,548
2009									
Grand total	188	1,105,744	59,849	194,173	157,859	411,881	37.25	−679	12,785

*In forest cover

GROWING STOCK IN FOREST AND TREE OUT OF FOREST

Growing stock of physiogeographic zone of north-east was computed to 443.09 Mm³ in 2007. It decreased from the value of 503.50 Mm³ reported in 2005 mainly due to shifting cultivation [17].

In the north-eastern states, in 2010, growing stock in forest and tree outside of forest (TOF) were recorded to be 932.165 Mm³ and 178.988 Mm³ respectively, which have increased over the reported value for 2007 mainly because of regeneration and afforestation programme (Table 3).

Higher growing stock of forest and TOF were recorded for Arunachal Pradesh followed by Assam, whereas lowest were in Sikkim. However, the north-eastern states contributed 20.72% and 11.56% of the total growing stock of forest and TOF of the country respectively, thereby contributing significantly to the growing stock in forest owing to large forest cover and high productivity of forest trees.

CARBON STOCK IN NORTH-EASTERN STATES

Carbon stock was reported to be 230.19 Mt in 2007 in physiogeographic zone of north-eastern India, which has decreased from a value of 261.05 Mt in 2005 [17]. Carbon stock varied in the different forest types, that is, montane wet temperate, sub-tropical broad-leaved, sub-tropical pine forest, and tropical deciduous in the north-eastern region. Aboveground biomass and carbon density in different forest types of north-eastern India are given in Table 4. Maximum aboveground biomass

Table 3 : North-eastern state-wise growing stock in forests and TOF (2010) [16]

State	Geographical area (km²)	Recorded forest area (km²)	Volume of growing stock (Mm³)		
			In forest	In TOF	Total
Arunachal Pradesh	83,743	51,540	492.689	74.516	567.205
Assam	78,438	26,832	173.494	41.336	214.830
Manipur	22,327	17,418	70.878	10.691	81.569
Meghalaya	22,429	9,496	45.411	20.964	66.375
Mizoram	21,081	16,717	68.042	9.392	77.434
Nagaland	16,579	9,222	40.955	12.681	53.636
Sikkim	7,096	5,841	18.832	2.017	20.849
Tripura	10,486	6,294	21.864	7.391	29.255
Total of the states of North-eastern India	262,179	143,360	932.165	178.988	1,111.153
Total of all states or UT of India	32,87,263	769,538	4,498.731	1,548.427	6,047.158
Percentage of north-eastern India	7.976	18.629	20.721	11.559	18.375

and carbon density are exhibited by sub-tropical pine forest plantation (382.23 Mg ha⁻¹ and 191.11 Mg C ha⁻¹) followed by sub-tropical broad-leaved hill forest of Manipur (344.70 Mg ha⁻¹ and 172.30 Mg C ha⁻¹) [18]. Minimum value of aboveground biomass and carbon density were recorded for tropical moist deciduous forest (20.35 Mg ha⁻¹ and 10.42 Mg C ha⁻¹). Lower value of carbon stock in the tropical moist deciduous forest dominated by *Dipterocarpus tuberculatus* was mainly due to over harvesting for timber and clearing the forest for shifting cultivation [19]. It shows that in the different forest ecosystems, the carbon stock in the forests depends on the age, species composition, and nutrient status of the soil.

According to IPCC [20] and Food and Agriculture Organization (FAO) [3], an average carbon density of 86 Mg ha⁻¹ and 81 Mg ha⁻¹, respectively, in the vegetation of world forests for the mid-1990s has been estimated. Carbon density was comparatively higher in plantation forest and ranged from 86.73 Mg C ha⁻¹ to 295.00 Mg C ha⁻¹ in pine

Table 4 : Aboveground biomass and carbon density in different forests of Manipur, north-eastern India

S. No.	Forest type	Altitude (m)	Age	Biomass (Mg ha^{-1})	Carbon density (Mg ha^{-1})	Location	Sources of data
1.	Montane wet temperate forest	1,800	30	199.23	99.61	Siroi hill, Manipur	[18]
2.	Sub-tropical broad-leaved hill forest	820	40–42	192.63	96.31	Maram, Manipur	[18]
3.	Sub-tropical broad-leaved hill forest	890	–	218.00	109.00	Bishnupur, Manipur	[11]
4.	Sub-tropical broad-leaved hill forest	950	–	344.70	172.30	Tamenglong, Manipur	[11]
5.	Sub-tropical pine forest plantation	910	16–40	382.23	191.11	Imphal valley	[18]
6.	Tropical moist deciduous forest	330	16	20.35	10.42	Moreh, Manipur	[19]
7.	Tropical semi-evergreen forest	295	–	323.90	161.90	Nongkhyllem, Meghalaya	[12]
8.	Tropical semi-evergreen forest	300	–	261.64	130.32	Barak valley, Assam	[33]

plantation in Manipur, whereas it was 203.2 Mg C ha^{-1} in the sal plantation in Meghalaya. Carbon density in pine forest of Manipur is higher than in the pine forest of Philippine [21].

Thus, it shows that plantation forests sequester more carbon in comparison to natural forests because of being better managed, whereas natural forests are generally subjected to heavy felling of trees and regenerated forest on abandoned shifting cultivation areas. Therefore, plantation of various indigenous tree species in the waste land, open forest, and abandoned shifting cultivation areas would be an appropriate strategy to enhance the carbon sequestration in the north-eastern region. Thus, sustainable management of both natural and plantation forests is essential to achieve sustainable development and to capture and store high level of carbon in soil and vegetation. The biomass (aboveground and belowground) and carbon stock for the north-east have been computed by Sheikh *et al.* [17] on the basis of inventory data of different forest types in the country. According to them, the north-eastern physiogeographic zone contributed 261.05 Mt C and 230.19 Mt C in the biomass of forest in 2005 and 2007 respectively, indicating decreasing trend in successive years.

RATE OF CARBON SEQUESTRATION

The data of carbon sequestration in the forest vegetation in the north-eastern region are very limited. The rate of carbon sequestration in selected forest types of Manipur has been given in Table 5. The rate of carbon sequestration was highest (14.80 Mg C ha^{-1} yr^{-1}) in sub-tropical broad-leaved hill forest and lowest (3.49 Mg C ha^{-1} yr^{-1}) in tropical moist deciduous forest of Manipur. In pine plantation forest, it was estimated to be 10.00 Mg C ha^{-1} yr^{-1} [13, 22].

The rate of carbon sequestration is highly variable and depends on soil composition, nutrient status of soil, age of tree, level of biotic disturbance, and other climatic factors in the different forest types in the region.

CARBON SEQUESTRATION IN SOIL

Soil is the largest pool of terrestrial organic carbon [23–24]. The data on soil organic carbon (SOC) is limited in the forest ecosystems of the north-eastern region [13–15]. Most of the carbon is sequestrated in the upper layer of the soil (0–30 cm). Soil organic carbon density varied

Table 5 : Rate of carbon sequestration in different forests of Manipur, north-eastern India

S. No.	Forest type	Altitude (m)	Age	Net biomass production (Mg ha^{-1} yr^{-1})	Carbon flux (Mg C ha^{-1} yr^{-1})	Location	Sources of data
1.	Montane wet temperate forest	1,600	30	23.98	12.00	Siroi hill, Manipur	[18]
2.	Sub-tropical broad-leaved hill forest	820	32–42	29.58	14.80	Maram, Manipur	[18]
3.	Sub-tropical pine forest plantation	810	16–40	19.96	10.00	Imphal valley	[18]
4.	Tropical moist deciduous forest	330	16	6.98	3.49	Moreh, Manipur	[19]

from a minimum of 22.56 Mg C ha^{-1} in pine plantation to a maximum of 62.79 Mg C ha^{-1} in sub-tropical mixed oak forest, Senapati, Manipur (Table 6).

The reported data on SOC density for north-eastern state show that Sikkim registered high density (42.33 Mg C ha^{-1}). Nagaland, Manipur, and Meghalaya recorded comparable 31.0–33.4 Mg C ha^{-1}. Assam and Tripura recorded significantly low average density (22.3–29.5 Mg C ha^{-1}) compared to the other north-eastern states [15]. SOC density ranged from 47.29 Mg C ha^{-1} in sal forest to 93.47 C ha^{-1} in Kail and Quercus stand in Himachal Pradesh [25] and varied between 40.3 Mg C ha^{-1} and 177.5 Mg C ha^{-1} in temperate forests of Garhwal Himalaya, India [26].

High value of SOC density in the forest ecosystems is governed by high rate of litter production and faster decomposition of litter [27]. SOC density is highly variable depending upon the climatic conditions ranging from alpine to tropical, topography, rainfall pattern, forest types in the north-eastern region. With the increase in elevation, SOC content also consistently increased in the different north-eastern states, that is high in Sikkim followed by Nagaland, Manipur, Meghalaya, and lowest in Assam and Tripura [15]. With the increase in altitude from mean sea level, temperature decreases at the rate of 1°C for each 166 m decrease [28]. The low temperature may favour high SOC content (2.99%) as in the case of Sikkim located at higher altitude and low SOC content in Tripura, which is at lower altitude with high temperature. This may be due to low rate of decomposition and mineralization owing to low temperature in the former. Similar trend in SOC density was reported by Wani *et al.* [25] and Sharma *et al.* [26].

Table 6 : Mean organic carbon content in the soils of different forests of Manipur, north-eastern India

S. No.	Forest type	Altitude (m)	Mg C ha^{-1}	Location	Sources of data
1.	Sub-tropical broad-leaved hill forest	820	27.73	Maram, Manipur	[13]
2.	Sub-tropical pine forest plantation	810	22.56	Imphal valley	[13]
3.	Tropical moist deciduous forest	330	48.03	Moreh, Manipur	[19]
4.	Sub-tropical mixed oak forest	998	55.02	Senapati, Manipur	[32]
5.	Sub-tropical mixed oak forest	1,294	62.74	Senapati, Manipur	[32]

The surface SOC stock of north-eastern states covering 64.71% of the total geographical area representing major land-use systems was reported recently [15]. Out of the total SOC stock of 339.82 Tg of north-eastern states, dense forest, open forest, plantation, and shifting cultivation contributed 43.31%, 8.82%, 4.76%, and 3.14% of soil carbon stock, respectively in the different states.

Thus, forest contributes more than 50% of the total soil carbon stock of the total geographic area represented by major land-use systems of the north-eastern region. It shows that forest soils in the north-eastern states sequester more carbon as compared to the other states of the country owing to high SOC density and high percentage of forest cover, thereby contributing significantly. The information on the soil carbon distributed in deeper layers in roots is lacking, which holds a great potential of carbon sequestration in the soil. Therefore, the role of soil carbon in the roots and carbon leaching in deeper layer need to be further investigated. The carbon must be fixed into long-lived forests; otherwise one may simply alter the size of fluxes in carbon cycle, not increasing the carbon sequestration.

SOIL CO_2 EMISSION FLUXES

Annual soil CO_2 emissions varied from 1.58 t CO_2 ha^{-1} yr^{-1} to 4.46 t CO_2 ha^{-1} yr^{-1} in different forests of Manipur (Table 7). Montane wet temperate and tropical moist deciduous forest exhibited high rate of soil CO_2 flux.

Emissions are highly variable depending on the soil carbon stock and climatic factors. A number of studies have reported that temperature was the single most important variable for predicting the soil CO_2 flux [29–31]. Soil CO_2 flux is strongly influenced by seasons and abiotic variables in the sub-tropical forests of Manipur [29].

The highest rate of CO_2 emissions (56.62 Mt yr^{-1}) has been reported from forest destruction in north-eastern physiogeographic zone followed

Table 7 : Annual CO_2 emissions from the soils in different forests in Manipur, north-eastern India

S. No.	Forest type	Mg CO_2 ha^{-1} yr^{-1}	Sources of data
1.	Montane wet temperate forest	4.46	[18]
2.	Sub-tropical broad-leaved hill forest	1.58	[29]
3.	Tropical moist deciduous forest	3.77	[18]
4.	Sub-tropical mixed oak forest	2.46	[32]

by eastern Himalaya (24.04 Mt yr^{-1}) and lowest (5.92 Mt yr^{-1}) from western plains [17].

FOREST MANAGEMENT PRACTICES IN NORTH-EASTERN INDIA

Forest resources are generally managed by adopting two types of management practices in the region: indigenous forest management and state forest management.

Indigenous Forest Management

Individual and community forests are managed by local organizations, traditional leaders, and district councils. They resolved the important issues of (i) rotation and allocation of forest blocks for shifting cultivation (Jhuming cultivation), (ii) protection of watershed and sacred forest, (iii) equitable distribution of forest products, and (iv) even dispute of forest lands. The government generally has no authority to regulate private forest lands, but provides incentives in the form of subsidies and technical assistance to increase and maintain the forest cover.

State Forest Management

The State Forest Department implements various forest conservation and development programmes through Joint Forest Management (JFM), Village Forest Management Committee (VFMC), and Forest Development Agency (FDA). Afforestation and reforestation programmes are being implemented by respective State Forest departments in consultation with the local communities under various forest management practices in north-eastern India. For example, selection of plant species for plantation, suitability of areas in the degraded forests and abandoned shifting cultivation areas, rotation and reducing harvest damage, and so on. However, these programmes are generally funded by national and international agencies.

More than 41% of the total forest cover falls under open forest besides shifting cultivation areas in the region. Therefore, the north-eastern states have great opportunities in reduction of CO_2 through afforestation and reforestation programmes in the large tract of open forest and shifting cultivation areas through funding mechanism of Global Environmental Facility (GEF), Clean Development Mechanism (CDM), and Reducing Emission from Deforestation and forest Degradation (REDD).

CONCLUSION

Forests in north-eastern India play an important role in carbon sequestration in soil and forest vegetation. Most of the north-eastern states occupy more than 77% of the forest cover, out of the total geographical areas of the respective states except Sikkim and Assam [16]. It comprises 7.98% of the total geographical area of the country, but accounts for nearly 25% of the total forest of the country.

However, there was a loss of 549 km^2 forest cover in the north-eastern region during 2007–09. Maximum reduction in forest cover was recorded in Manipur (190 km^2) followed by Nagaland (140 km^2), though, it declined in all the states except Sikkim. It is a matter of great concern for the states and requires great efforts in increasing the forest cover to combat the reduction of CO_2.

Growing stock was recorded to be 932.165 Mm^3 in 2010 for the north-eastern region and contributed about 20.72% of the total growing stock of the country. Carbon stock varied from 10.42 Mg C ha^{-1} to 172.30 Mg C ha^{-1} and the rate of carbon sequestration varied from 3.49 Mg C ha^{-1} to 14.80 Mg C ha^{-1} in the different forest ecosystems of north-eastern region and is highly variable depending upon the types of forests, age, species composition, and level of biotic disturbances. However, plantations sequester high percentage of carbon in soil and tree biomass in the region devoid of biotic disturbances and better management practices.

Carbon stock in the forest of north-eastern physiogeographic zone was reported to be 261.05 Mt in 2005 which decreased to 230.19 Mt in 2007. The total SOC stock of soils of the north-eastern states covering 64.71% of the total geographic area under different land-use systems was 339.82 Tg, out of which forest contributed more than 50% of the total SOC stock. SOC density ranged from a minimum of 22.56 Mg ha^{-1} in pine plantation to a maximum of 62.24 Mg ha^{-1} in sub-tropical forests of Manipur. However, SOC density was reported to be highest (43.33 Mg C ha^{-1}) for Sikkim and lowest in Tripura (22.3–29.5 Mg ha^{-1}). It increases with the increase in altitudes and is reported high in Sikkim and low in Tripura depending upon the temperature, which is low in the former and high in the latter. SOC density also depends upon litter production and faster decomposition of litter.

Annual soil CO_2 emission flux ranged from 1.58 Mg CO_2 ha^{-1} yr^{-1} to 4.46 Mg CO_2 ha^{-1} yr^{-1} in different forest ecosystems of Manipur, which is highly variable depending upon soil organic carbon stock and abiotic variable, especially temperature and rainfall.

Therefore, forests in north-eastern India play a vital role in carbon sequestration in soil–vegetation systems not only because of large forest cover, but also due to high productive potential of forest ecosystems. GEF, CDM, and REDD provide funds to increase carbon stock in forests and to reduce emission CO_2 quota in the developing countries. Therefore, a unique opportunity for north-eastern states to engage in mitigation of CO_2 through mass scale reforestation and afforestation programme in large tract of open forests and abandoned shifting cultivation areas of the region exists.

The current information on carbon stock and rate of sequestration in the forest ecosystems would be useful to devise appropriate forest management policies to increase the carbon stock in the forests of north-eastern region and contribute significantly in the reduction of CO_2 to the atmosphere at regional and national level and to mitigate climate change in future.

Summary

Carbon management in forests is one of the important agenda in India to reduce emission of CO_2 and to mitigate global climate change. North-eastern region of the country comprises Arunachal Pradesh, Assam, Manipur, Meghalaya, Mizoram, Nagaland, Sikkim, Tripura and the total forest cover is 173,219 km^2, which is 66.07% of its geographical area in comparison to the national forest cover of 21.5% of the total geographic area of India. The north-eastern region constitutes only 7.98% of the geographical area of the country, but accounts for nearly 25% of the forest cover of the country. There was a loss of 549 km^2 of forest cover during 2007–09 in the north-eastern states mainly due to shifting cultivation. The chapter discusses the carbon sequestration in soil, aboveground biomass, and soil CO_2 flux in different forest ecosystems of the north-eastern region to formulate strategies to optimize carbon sequestration in the forests of the region.

REFERENCES

1. Stern, N. 2006. *Review: The Economics of Climate Change.* HM Treasury, UK: Cambridge University Press

2. Brown, S., J. Sathaye, M. Cannel, and P. Kaeppi. 1995. Management of forests for mitigation of greenhouse gas emission. *Climate Change. Impact Adaptation and Mitigation of Climate Change: Scientific–technical Analysis.* Contribution of Working Group II to the Second Assessment Report of the Intergovernmental Panel for Climate Change, R. T. Watson, M. C. Zinyowera, and R.H. Moss (eds), pp. 775–797. Cambridge: Cambridge University Press

3. FAO. 2005. Global forest resources assessment: Progress towards sustainable forest management. *FAO Forest Paper Rome* 147

4. Ravindranath, N. H., R. K. Chaturvedi, and Indu. K. Murthy. 2009. Forest conservation, afforestation and reforestration in India. Implication for forest carbon stock. *Current Science* 95:216–222

5. Intergovernmental Panel for Climate Change (IPCC). 2007. *The IPCC Fourth Assessment Report.* Cambridge: Cambridge University Press

6. Ravindranath, N. H., B. S. Somasekhar, and Madhav Gadgil. 1997. Carbon flow in Indian forests. *Climate Changes* 35:297–320

7. Lal, M. and R. Singh. 2000. Carbon sequestration potential of Indian forests. *Environmental Monitoring and Assessment* 60:315–327

8. Chhabra, A., S. Palria, and V. K. Dadhwal. 2002. Soil organic carbon pool in Indian forests. *Forest Ecology and Management* 173:187–199.

9. Dadhwal, V. K., S. Singh, and P. Patil. 2009. Assessment of phytomass carbon pools in forest ecosystems in India. *National Natural Resource Management System Bulletin* 33:41–57

10. Devagiri, G. M., S. Money, S. Singh, V. K. Dadhawal, P. Patil, A. Khaplek, A. S. Devakumar, and S. Hubballi. 2013. Assessment of aboveground biomass and carbon pool in different vegetation types of south-western part of Karnataka, India using spectral modeling. *Tropical Ecology* 54(2):149–165

11. Thokchom, A. and P. S. Yadava. 2013. Biomass and carbon stock assessment in sub-tropical forests of Manipur, north-east India. *International Journal of Ecology and Environmental Science* 32(2):107–113

12. Baishya, R., S. K. Barik, and K. Upadhya. 2009. Distribution pattern of aboveground biomass in natural and plantation forest of humid tropics in north-east India. *Tropical Ecology* 50:295–304

13. Yadava, P. S. 2010. Soil and vegetation carbon pool and sequestration in the forest ecosystem of Manipur, N.E. India. In *CO$_2$ Sequestration Technology on Clean Energy*, S. Z. Qasim and Malti Goel (eds), pp. 163–170. Delhi: Daya Publication House

14. Bhattacharya, T., D. Sarkar, D. K. Pal, C. Mandal, C. Baruah, D. Telpande, and P. H. Vaidhya. 2010. Soil information system for resource management – Tripura as a case study. *Current Science* 99:1208–1217

15. Choudhury, B. U., K. P. Mohapatra, Anup Das, Pratibha T. Das, L. Nongkhlaw, R. A. Fiyaz, S. V. Ngachan, S. Hazarika, D. J. Rajkhowa, and G. C. Munda. 2013. Spatial variability in distribution of organic carbon stocks in the soils of north-east India. *Current Science* 104:604–614

16. Forest Survey of India (FSI). 2011. *State of Forest Report.* Ministry of Environment and Forest, Government of India

17. Sheikh, M. A., M. Kumar, R. W. Bussman, and N. P. Todaria. 2011. Forest carbon stock and fluxes in physiogeographic zones of India. *Carbon Balance and Management* 6:1–15

18. Yadava, P. S. 2012. The role of forests in carbon sequestration and climatic change in north-east India with special reference to Manipur. *Souvenier Article. National Seminar on Carbon Sequestration in Terrestrial and Aquatic Ecosystem,* Manipur University, 14–27

19. Devi, L. S. and P. S. Yadava. 2009. Aboveground biomass and net primary production and semi-evergreen tropical forest of Manipur north-east, India. *Journal of Forestry Research* 20(2):151–155

20. Intergovernmental Panel for Climate Change (IPCC). 2000. Land-use, land-use change and forestry. *IPCC Special Report on Land-use, Land-use Change and Forestry.* London: Cambridge University Press

21. Lasco, R. D. and F. B. Pulhin. 2003. Phillipine forest ecosystems of climate change: Carbon stock, rate of sequestration of Kyoto Protocol. *Annual of Tropical Research* 25:37–51

22. Yadava, P. S. and L. S. Devi. 2009. Carbon stock and rate of carbon sequestration in dipterocarpus forests of Manipur, N.E. India. *Proceedings of XII World Forestry Congress,* Buenos Aires, Argentina

23. Schlesinger, W. H. 1990. Evidence from cheonosequence studies for a low carbon storage potential of soils. *Nature* 348:232–234

24. Lal, R. 2004. Carbon sequestration in India. *Climate Change* 65:277–296

25. Wani, A. A., P. K. Joshi, O. Singh, and R. Pandey. 2012. Carbon sequestration potential of Indian forestry land-use system – A review. *Nature and Science* 10:78–85

26. Sharma, C. M., S. Gairola, N. P. Baduni, S. K. Ghildiyal, and S. Suyal. 2011. Variation in carbon stocks on the different slope aspects in seven major forest types of temperate region of Garhwal Himalaya, India. *J. Bioscience* 36:701–708

27. Devi, A. S. and P. S Yadava. 2007. Wood and leaf decomposition of *dipterocarpus tubuculatus roxb.* in a tropical deciduous forest ecosystem of Manipur, North-east India. *Current Science* 93(2):243–246

28. Lal, R. and M. K. Shukla. 2004. *Principles of Soil Physics.* New York, Basel: Marcel Dekker

29. Devi, N. B. and P. S. Yadava. 2009. Emission of CO_2 from the soil and immobilization of carbon in microbes in a sub-tropical mixed oak forest ecosystem, Manipur, N.E. India. *Current Science* 96(12):1630–1637

30. Bijracharya, R. M., R. Lal, and J. M. Kimbe. 2002. Diurnal and seasonal CO_2 C flux from soil as related to erosion phases in Central Ohio. *Soil Science Society American Journal* 64:286–293

31. Rastogi, M., S. Singh, and H. Pathak. 2002. Emission of carbon dioxide from the soil. *Current Science* 82:510–517

32. Devi, Y. J., P. S. Yadava, and B. K. Dutta. 2012. Seasonal changes in the CO_2 emission of the soil in the sub-tropical forest of Sauntak Molnom, Senapati District, Manipur. *Proceeding Souvenier Article. National Seminar on Carbon Sequestration in Terrestrial and Aquatic Ecosystem*, Manipur University, 111

33. Borah, N., A. J. Nath, and A. K. Das. 2012. Aboveground biomass and carbon stock in two reserve forest of Cachar District, Assam, India. *Proceeding Souvenier Article. National Seminar on Carbon Sequestration in Terrestrial and Aquatic Ecosystem*, Manipur University, 107

Section III
CO$_2$ FIXATION AND UTILIZATION

CHAPTER **11**

Prospects in Biomimetic Carbon Sequestration

Shazia Faridi and T. Satyanarayana
Department of Microbiology, University of Delhi, South Campus,
New Delhi-110021
E-mail: tsnarayana@gmail.com

INTRODUCTION

The adverse climatic changes occurring all over the world indicate that global warming has reached an alarming level due to increasing concentrations of greenhouse gases (GHGs) in the atmosphere. It has also created pressure to develop strategies to reduce these changes [1]. Carbon dioxide (CO_2) is the major contributor to greenhouse effect in terms of both emission and its climate altering potential. The atmospheric CO_2 concentration has increased from a pre-industrial level of 280 parts per million (ppm) to 390 ppm at present. Several attempts are being made to develop alternatives to fossil fuels, such as using renewable energy sources. Fossil fuels will probably continue to rule the next few decades. Thus, there is an urgent need to address the increasing level of CO_2 in the atmosphere. Towards this end, various carbon sequestration technologies are being assessed for mitigating CO_2 levels in the atmosphere [2].

Physical means of CO_2 storage (geological and ocean sequestration) are associated with high costs and risk of leakage. In this context, the use of biological sequestration (in short bio-sequestration) processes is considered viable and promising for reducing emissions of CO_2 in the atmosphere. Bio-sequestration is the capture of CO_2 from the atmosphere by the biological processes. Bio-sequestration of CO_2 may be by enhanced photosynthesis by increasing soil carbon storage, or by using heterotrophic bacteria and their enzymes [3]. This chapter is aimed at summarizing the developments in carbon bio-sequestration.

BIO-SEQUESTRATION THROUGH PHOTOSYNTHESIS

Preserving and adding to the world's forest canopy is the cheap and easily manageable natural means for minimizing the impact of global warming. Kindermann *et al.* [4] reported that by avoiding deforestation in tropical regions, around 2.8 billion tonnes of CO_2 emissions could be reduced per year. Increasing the earth's percentage of plants which have C_4 carbon fixation mechanism will also contribute to bringing down atmospheric CO_2 level. The efforts are also underway to increase the photosynthetic efficiency of RuBisCO genes to increase the catalytic and/ or oxygenation activity of the enzyme [5].

Another way by which plants can act as carbon sinks is by the use of bioenergy crops, thereby minimizing GHG emissions from fossil fuels. Bioenergy crops provide a carbon neutral energy source. Today sugarcane, oil crops, and cereals, particularly maize and wheat, make the largest contribution to bioethanol. Globally appropriate forest policies could increase the amount of carbon sequestered in terrestrial biomass by up to 100 Gt or up to 2 Gt per year [6].

CARBON CAPTURE BY SOIL

Soil contains three times more carbon than the amounts stored in living forms. Long-term storage of carbon in soil can be attained when carbon from aboveground biomass enters the pool of soil organic carbon (SOC) or soil inorganic carbon (SIC).

One of the approaches to sequester carbon in soil involves development of a carbon farm which would be devoted to growing biomass for the production of phytoliths or biochar. Phytoliths are microscopic spherical particles of silicon which can store carbon for thousands of years [7], while biochar is charcoal which is created by pyrolysis of biomass. *Panicum virgatum*, a perennial switch grass which is valuable in biofuel production and soil conservation, has increased soil organic carbon at 0–12 inches and 0–47 inches at the rates of 0.5 and 1.3 tonnes carbon acre^{-1} yr^{-1} (equivalent to 1.8 and 4.7 tonnes CO_2 acre^{-1} yr^{-1}), respectively [8].

MICROBES IN CARBON FIXATION

The ability of microbes, especially prokaryotes to recycle the essential elements that make up the cell, significantly affects the environment. The global carbon cycle is profoundly dependent on microbes. Microbes are considered to be capable of capturing CO_2.

Role of Microbes in Photosynthesis

In addition to plants, photosynthetic microbes such as algae and autotrophic bacteria (cyanobacteria and anoxyphotoautotrophs) also play an important role in the fixation of CO_2. They play an important role in carbon cycle by converting atmospheric CO_2 into organic matter. Cyanobacteria fix CO_2 and produce oxygen during photosynthesis, and they make a very large contribution to the carbon and oxygen cycles. Cyanobacteria can be developed as a microbial cell factory for converting atmospheric CO_2 into beneficial products by using solar energy.

ALGAL SEQUESTRATION OF CO_2

Microalgae are capable of carrying out photosynthesis using free CO_2 and bicarbonate ions as a source of inorganic carbon. The rapid growth rates of algae and their capacity to grow practically in any kind of environment makes them attractive for capturing CO_2 emitted by power plants and other industrial sources worldwide. Algae are known to thrive at high concentrations of CO_2 and nitrogen dioxide (NO_2). Kubler *et al.* [9] studied the effect of elevated CO_2 levels on the seaweed *Lomentaria articulata*, and found that twice the ambient CO_2 concentration had significantly affected daily net carbon gain and total wet biomass production rates; 52% and 314% greater than they were under ambient CO_2 conditions, respectively. The microalgae such as *Chlorella* grow much better at 100,000 ppm of CO_2 than in the ambient air [10]. It can, therefore, be concluded that a pollutant from power plants (flue gases containing CO_2) can act as a nutrient for the algae. Thus, flue gases from power plant emissions such as from coal-based power plants can be fed to algal production facilities to significantly increase productivity resulting in capture and storage of CO_2 in the biomass. For making this process cost-effective, algal biomass generated after sequestration can be used for the production of biofuel, food supplements for humans, pharmaceuticals and enzymes, animal feed, and electricity generation upon combustion directly or by anaerobically transforming the algal biomass to CH_4.

Seambiotic is the first company in the world that cultivated algae using flue gas from coal burning power station.

HETEROTROPHIC BACTERIA IN CARBON SEQUESTRATION

Wood *et al.* [11] proposed that CO_2 is reduced during the fermentation of glycerol by certain representative genera of heterotrophic bacteria such as propionic acid bacteria, *Propionibacterium*, *Escherichia*, and *Citrobacter*, and further showed that CO_2 and pyruvate combine to form oxaloacetate. This pathway of oxaloacetate formation using CO_2 and pyruvate can be exploited for carbon capturing using heterotrophic bacteria involving carbonic anhydrase (CA). The ability of CA to convert CO_2 into bicarbonate may be utilized by carboxylases such as phosphoenolpyruvate (PEP) carboxylase and pyruvate carboxylase to form oxaloacetate [12]. Such anapleurotic pathway exists in organisms to compensate the loss of oxaloacetate siphoned off for the synthesis of amino acids of the aspartate family.

Thus, heterotrophic bacteria can be used to sequester carbon and in turn produce oxaloacetate and amino acids by providing CO_2. The production of amino acids such as glutamic acid and lysine by *Corynebacterium glutamicum* is well known. This bacterium contains the enzymes, PEP carboxylase and pyruvate carboxylase and PEP pyruvate-oxaloacetate node [13], and, thus, fixing carbon in the form of amino acids. In elevated CO_2 conditions, increased supply of bicarbonate by the action of carbonic anhydrases, PEP carboxylase and pyruvate carboxylase activities make the conditions favourable for lysine production [14].

NON-PHOTOSYNTHETIC CO_2 FIXATION BY HETEROTROPHIC BACTERIA

Methane (CH_4) is the main component of natural gas. The biological conversion of CO_2 to organic compounds like CH_4 is an attractive method for the sequestration of CO_2. Microorganisms such as hydrogenotrophic methanogens can very efficiently convert CO_2 to CH_4 as they utilize CO_2 as a source of carbon and hydrogen as a reducing agent. A part of CO_2 reacts with hydrogen to produce CH_4, generating electrochemical gradient across a membrane, which is then used to generate adenosine triphosphate (ATP) through chemiosmosis ($CO_2 + 4H_2 \rightarrow CH_4 + 2H_2$). These methanogens inhabit anaerobic environments that contain CO_2, acetate, and low sulphate concentrations. They grow in the temperature range between 9°C and 110°C.

The bioconversion of CO_2 to CH_4 offers several advantages over chemical conversion. First, it requires much lower energy for reduction of CO_2. Second, burning CH_4 produced by these microbes do not contribute to atmospheric CO_2 as these microbes use CO_2 from air to form CH_4, and hence, burning the fuel will just return the extracted gas. CH_4 from microbes is more eco-friendly than other biofuels produced from feedstock such as corn ethanol. It does not compete with food production and other agricultural resources such as land, irrigation, and fertilizers.

Hydrogenotrophic methanogens are known to inhabit oil reservoirs universally. Thus, these methanogens can be exploited for *in-situ-*conversion of CO_2 into CH_4. This conversion process has not only the potential of reducing CO_2 emission but also the potential to reproduce natural gas deposit in reservoirs. Lee *et al.* [15] analysed biological conversion of CO_2 to CH_4 using hydrogenotrophic methanogens in a fixed bed reactor. They concluded that methanogens have the potential to be effective in converting CO_2 to CH_4 with a conversion rate of 100% at 3.8 hours retention time.

The successful application of naturally occurring methanogens to produce CH_4 using CO_2 would allow the conversion of sub-economic (high CO_2) natural gas deposits into pure and economical CH_4 deposits, and would also open up this carbon capture and storage (CCS) technology to widespread application by power generation, chemical, petroleum, and other industries. This is, therefore, an area of high-potential research that can be explored and possibly exploited.

ENZYMATIC CARBON CAPTURE

Carbonic Anhydrase in Carbon Capture

The biomimetic use of carbonic anhydrase provides an attractive way to capture CO_2 in an energy-efficient way directly from industrial flue gas, producing a raw material that can be sequestered underground or turned into useful substances such as baking soda, chalk, or limestone. This enzyme is an effective biocatalyst and nature's way to efficiently convert CO_2 to bicarbonate. Carbonic anhydrases (CAs, EC 4.2.1.1) are zinc containing metalloenzymes which catalyse the reversible hydration of CO_2.

$$CO_{2\ (aq)} + H_2O \rightarrow HCO_3^- + H^+$$

These are among the fastest enzymes known. Human isozyme Human Carbonic Anhydrase Isozyme (HCAII) is the fastest known CA, each molecule of which can hydrate at least 1.4×10^6 molecules of CO_2 per second [16]. The reaction occurs at or near the diffusion-controlled limit for the encounter rate of enzyme and CO_2.

$$E\text{-}ZnH_2O \rightarrow E\text{-}ZnOH^- + H^+$$
$$E\text{-}ZnOH^- + CO_{2\ (aq)} \rightarrow E\text{-}ZnHCO_3^-$$
$$E\text{-}ZnHCO_3^- + H_2O \rightarrow E\text{-}ZnH_2O + HCO_3^-$$

Microbial carbonic anhydrases play an essential role in various physiological processes including respiration, supply of bicarbonate to various metabolic pathways, ion transport, acid–base regulation, and biomineralization. They are also an important component of carbon concentrating mechanism in various photosynthetic organisms, plants to algae, and cyanobacteria.

Process of Carbon Capture by CA

CO_2 Solutions Inc. describes the use of CA as an onsite scrubber of CO_2 [17]. The capture of CO_2 directly from flue gases by carbonic anhydrase and its separation from other flue gases requires a bioreactor containing a matrix onto which CA is immobilized and which acts as a solid support for the enzyme. The process is shown in Figure 1. The use of carbonic anhydrase as a biocatalyst in the process brings down the cost normally associated with other carbon capture methods.

Bhattacharya *et al.* [18] used a variation of this process in which the enzyme was immobilized by covalent grafting on silica coated porous steel by immobilizing the enzyme and water was sprayed down through the flue gas. Spraying water speeded up the dissolution of CO_2. The optimal CO_2 reduction was obtained with the reactor design that allows the horizontal inflow and outflow of the emission gas (at 60°C) and water sprayed from vertical position. Iliuta and Larachi [19] showed that gas–liquid and liquid–solid mass transfer exchange mechanisms could considerably modify the CO_2 hydration kinetics for packed bed reactors. Integrating immobilized enzyme absorption with ion-exchange resin resulted in the removal of excess enzyme which could inhibit bicarbonate ions [20]. Based on the studies, they developed a three-phase reactor, having HCAII enzyme, immobilized in the longitudinal channels washcoat of a post-combustion column, in which aqueous slurry containing resin exchange beads was flowed as shown in Figure 2 [20].

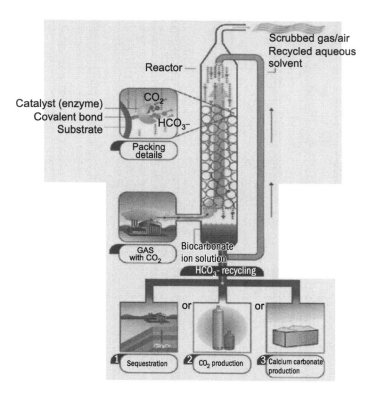

Figure 1 *An illustration of CO_2 capture unit developed by CO_2 Solutions Inc. (www.co2solutions.com/en/the-process) [17]*

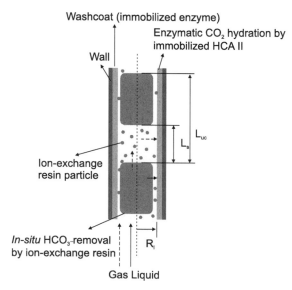

Figure 2 *A model of the three-phase reactor developed by Iliuta and Larachi [19]*

Another process of CA-based CO_2 capture was initiated by the National Aeronautics and Space Administration (NASA) for the purification of the ambient atmosphere of confined inhabited cabins, by employing thin aqueous films carrying dissolved CA to capture CO_2 [21–22]. The concentration of CO_2 in such atmospheres was 0.1% or less. The enzyme selectively allowed CO_2 to diffuse in the ratio of 1400 : 1 by comparison with nitrogen and 866 : 1 by comparison with oxygen [22]. A schematic illustration of the membrane sandwich involved is shown in Figure 3.

The core of the liquid membrane comprises a thin (for example, 330 μm thick) layer of enzymatic solution in an aqueous phosphate buffer, squeezed in between two microporous hydrophobic polypropylene membranes, retained by thin metal grids to protect the liquid membrane's thickness and rigidity. The CO_2 from the atmosphere purifies spontaneously and dissolves inside the liquid membrane on one face of the membrane. It diffuses across the liquid membrane and evaporates out the other liquid membrane on the opposite face, either in vacuum or in a carrier gas. Matsuyama and colleagues used such enzymatic liquid films with gases containing up to 15% CO_2, closely representing the CO_2 concentration in flue gas [23–24].

Another CO_2 capture technology using CA was developed by Carbozyme Company which is illustrated in Figure 4. It is based on hollow microporous propylene microfibre, separated by control separators which are made of thin oxide powders. The whole system bathes in an excess of aqueous CA solution [25].

The enzyme is immobilized directly on the external faces of the microfibre. Water vapour under moderate vacuum (15 kPa) was used as the sweeping gas at a low flow rate in the released microfibres. The CO_2

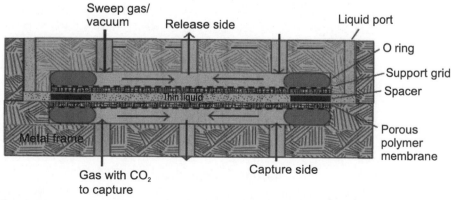

Figure 3   *An illustration of thin liquid membrane system used by NASA [22]*

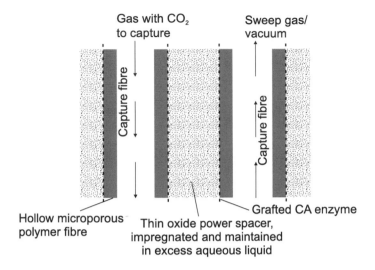

Figure 4   *An illustration of hollow microporous membrane-based CO_2 capture, such as that used by Carbozyme Inc. [25]*

content in the sweeping gas reached ≈95%, for a flue gas containing ≈15% CO_2 [25].

Several improvements in designs have been made, concerning the microfibre network geometry, the nature of hollow microfibres, and the variety of CA used. This technique was found to be efficient for a flue gas carrying 0.05% to 40% CO_2 at 15°C to 85°C with a particular γ-CA isozyme [26]. These bioreactors operate under harsh conditions (temperature of flue gases ranges from 50°C to over 125°C, presence of high concentration of organic amines and sulphur and nitrogen oxides). There is, therefore, a need to look for CA that remains active under such conditions. One of the main concerns related to the use of CA under such conditions is the possible effects of components present in the flue gas on the activity of enzyme. Bond et al. [27] have investigated the effects of SO_4^{2-} and NO_3^- ions on CA by directly supplementing these to the aqueous solution at 0°C to 5°C. They showed that with less than 0.1 M ion concentration, the effect was negligible. A γ-class CA purified from *Methanosarcina thermophila* (CAM) is optimally active at 55°C, with a k_{cat} value of 10^5/s. CAM is a comparatively stable CA that shows 50% residual activity after 15 minutes at 70°C and is inactivated at 75°C [28]. A γ-class CA has been studied from *Methanobacterium* (CAB) which is more thermostable than CAM, showing 50% residual activity after 15 minutes at 85°C and gets inactivated at 90°C, but has a lower k_{cat} value than (CAB) of 10^4/s [29].

Several other heat-stable CAs for extraction of CO_2 from CO_2-containing media had been reported. A detailed study on a thermophillic γ-class CA from *Bacillus clausii* has also been done. This CA exhibits higher thermostability than CAM with 17% residual activity after 15 minutes at 80°C in 1 M sodium bicarbonate at pH 8.05. A 0.6 g/l of this enzyme is capable of extracting greater than 99% of CO_2 from a 15% CO_2 gas stream against only 33% removal without CA [30]. Another approach to improve the activity and stability of CA under the non-natural conditions could be through protein engineering. Cordexis Company has done four rounds of directed evolution on a CA from a mesophillic organism which increased the enzyme stability with a residual activity of greater than 40% at 82°C in 5 M MDEA pH 11.8 and the half-life ($t_{1/2}$) increased by five times to 20 hours in 4.2 M MDEA at 75°C [31–32].

Immobilization of CA on various matrices is the most common method of stabilizing the enzyme in order to increase its reusability and to limit exposure to denaturing conditions for the accelerated CO_2 capture under relevant process conditions.

Different chitosan and sodium alginate-based materials were checked for immobilizing whole cells of *Bacillus pumilus* (extracellular CA producer), and these were found to improve the associated CA activity in comparison with the free cells [33]. Wanjari *et al.* [34] reported enhanced precipitation of $CaCO_3$ with purified CA from *Bacillus pumilus* adsorbed on chitosan beads as compared to the free enzyme. Prabhu *et al.* [35] studied the kinetics of CA immobilized in chitosan activated alumina-carbon composite beads, and observed that the K_m and V_{max} values were 10.35 mM and 0.99 μmole $min^{-1}ml^{-1}$. More than two-fold increase in $CaCO_3$ precipitation in a period of 5 minutes was attained, when purified CA from *Pseudomonas fragi* was immobilized by adsorption on chitosan beads in comparison with the free enzyme [36–37]. The kinetics of CA immobilized on ordered mesoporous aluminosilicate, which provides a high surface area and pore diameter, have been investigated by Wanjari *et al.* [38]. The K_m, V_{max}, and k_{cat} values for immobilized enzyme were 0.158 mM, 2.307 μ mole $min^{-1}ml^{-1}$, and 1.9 s^{-1}, and these for free CA were 0.876 mM, 0.936 μ mole $min^{-1}ml^{-1}$, and 2.3 s^{-1}, respectively. The immobilized CA exhibited CO_2 sequestration efficiency of 16.14 mg of $CaCO_3$/mg of CA compared to 33.08 mg of $CaCO_3$/mg of CA of free CA. Immobilized CA retained better stability and almost 67% of its initial activity even after six cycles [39]. The entrapped CA efficiently converts CO_2 to bicarbonate and/or carbonate that gets transformed into calcite on reacting with Ca^{2+} ions. Yadav *et al.* [40] calculated apparent Michaelis constants K_m and V_{max} for the precipitation of $CaCO_3$ using CA immobilized on silylated chitosan beads. A covalent coupling method was also developed to graft CA enzymes onto silica nanoparticles made by spray pyrolysis. Yadav *et al.* [41] studied the immobilization of CA by single or multiple attachments to polymers, themselves deposited onto Fe_3O_4 magnetic aggregates. The immobilized CA has longer storage stability than the free enzyme and retained 50% of its initial activity up to 30 days. They also developed core-shell single enzyme nanoparticles (SEN-CA) by covering the CA surface with a thin layer of chitosan, which showed an improved stability in comparison with the free enzyme [41]. Besides adsorption and covalent grafting, enzymes can also be efficiently entrapped within porous supports like polyurethane foam [42]. Polyurethane immobilized CA could be used without any activity loss in aqueous media for seven successive CO_2 capture tests and the optimum operational temperature was in a range from 35°C to 45°C. CA from *Pseudomonas fragi, Micrococcus lylae,* and *Micrococcus luteus* were compared with commercial bovine CA (BCA) by Sharma and Bhattacharya [43].

The extraction and purification of enzyme from wild-type microbes is a costly affair. Cloning of the gene coding for the particular enzyme provides an economic way for producing large quantity of that protein. Most CAs have been cloned from various pathogenic organisms, a very few reports are available from non-pathogenic microbes. Premkumar *et al.* [44] cloned α-type carbonic anhydrase (Dca) which is associated with the plasma membrane of the extremely salt-tolerant, unicellular, green alga *Dunaliella salina.* The CA is active over broad salinity of 0–4 M NaCl. Kaur *et al.* [45] cloned a putative γ-CA encoding gene of *Azospirillum brasilense* and over expressed it in *E. coli.* Its expression was induced in stationary phase and in the presence of high $CO_2.$ This CA is co-transcribed with the N-acetyl-γ-glutamate-phosphate reductase suggesting a possible link between arginine metabolism and an unknown CO_2-dependent metabolic process that utilizes this CA. An α-type CA of *Neisseria gonorrhoeae* was cloned and overexpressed in *E. coli.* The recombinant CA, in purified and non-purified crude form, shows comparable CO_2 hydration activity to commercial BCA and considerably promoted formation of solid $CaCO_3$ [46]. Recently a full length open reading frame of marine diatom encoding a CA gene from *Thalassiosira weissflogii* has been cloned and functionally expressed in *E. coli* using pTWIN2 expression vector [47].

Bicarbonate produced after CO_2 sequestration process finds many applications. Chi *et al.* [48] successfully utilized the bicarbonates produced from carbon capture as a feedstock for algae or cyanobacteria culture. The carbonate regenerated by the culture process was used as an absorbent to capture more CO_2. This process would significantly cut down the cost of the carbon capture process as transport of the aqueous bicarbonate solution costs far less than that of compressed CO_2, and also using bicarbonate would provide a superior alternative for CO_2 delivery to raise an algal culture system. In addition to this, it does not require additional energy for carbonate regeneration.

BIO-MINERALIZATION USING CARBONIC ANHYDRASE

Mineralization of CO_2 is a process that occurs naturally and is responsible for large amount of carbonate minerals such as calcite, aragonite, dolomite, and dolomitic limestone present on the earth's surface. The process is based on the reaction of CO_2 with metal oxides (of calcium and magnesium) bearing materials to produce insoluble carbonates. This natural process is termed silicate weathering. It uses naturally occurring minerals such as wollastonite ($CaSiO_3$), serpentine ($Mg_3Si_2O_5(OH)_4$), and olivine (Mg_2SiO_4), and consumes atmospheric CO_2. Weathering of silicates occurs in both freshwater and salt water as the waters have dissolved CO_2 that exists in equilibrium with dissolved CO_2, HCO_3^-, and CO_3^{2-}. The pH governs the balance of these three components, but the acidic components weather silicate minerals, releasing the divalent cations which react with bicarbonate to form mineral carbonates [49] as shown below:

Gaseous CO_2 dissolves rapidly in water to produce a loosely hydrated aqueous form, Equation (1).

$$CO_2 \text{ (g)} \rightarrow CO_2 \text{ (aq)} \tag{1}$$

The aqueous CO_2 then reacts with water to form carbonic acid, Equation (2).

$$CO_2\text{(aq)} + H_2O \rightarrow H_2CO_3 \tag{2}$$

In the next step, carbonic acid dissociates into bicarbonate and carbonate ions, Equations (3)–(4).

$$H_2CO_3 \rightarrow HCO_3^- \tag{3}$$

$$HCO_3^- + OH^- \rightarrow CO_3^{2-} + H_2O \tag{4}$$

Finally carbonate ions, in the presence of divalent metal cations, get precipitated as mineral carbonate, Equation (5).

$$CO_3^{2-} + Ca^{2+} \rightarrow CaCO_3\downarrow \tag{5}$$

Among the entire reactions for the mineralization of CO_2, Equation (2), the conversion of CO_2 to bicarbonate is the slowest, and, hence, the rate limiting step. The equilibrium constants for Equations (2) and (3) are 2.6 $\times 10^{-3}$ and 1.7×10^{-4}, respectively [50]. Moreover, Equations (3) and (4) are fast as the rates are diffusion controlled [51] and aqueous CO_2 in water can be transformed into H_2CO_3, HCO_3^- and CO_3^{2-}, depending on the pH. Carbonates have a tendency to get dissolved at low pH instead of getting precipitated. The hydration of CO_2 can be made faster under ambient conditions in the presence of an enzyme carbonic anhydrase that catalyses this rate limiting step awfully fast with a k_{cat} ranging from 10^5–10^7 s^{-1} [50, 52] This enzyme speeds up the hydration of CO_2 dramatically provided the pH is above the pK_a of CO_2/HCO_3^- equilibrium. The most active CA hydrates CO_2 at rates as high as $k_{cat} = 10^6$ s^{-1} or a million times a second. Thus, by employing CA at mildly basic pH (pH > 8), the entire process of mineralization of CO_2 can be made faster and thus this enzyme plays a vital role in the sequestration of CO_2 [50, 52]. The NMTI (New Mexico Technology Institute) and EPRI (Electric Power Research Institute) of America attempted CO_2 storage in the flue gas of a coal-fired power system using CA [53]. Interestingly, waste from oil industry such as metallurgical slags, lignite ashes, or chemical brine rejects [54] contain a significant concentration of such cations and could be used for carbonation.

Carbon dioxide sequestration in the form of mineral carbonation offers several advantages like the carbonates formed are environmentally safe and stable over geological time scales as they are fixed permanently, and, hence, there is no issue of future leakage of CO_2 from them. Raw materials for fixing CO_2 are abundant. And the technology does not require the concentration and transportation of CO_2 for ground or ocean storage. The process of mineral carbonation is thermodynamically favourable as the produced carbonates represent a lower energy state than CO_2.

The process has got the potential to be economically viable as the carbonates can be used for many applications which may further compensate its costs. Calcium carbonate has its main application in construction either as a building material (cement) or limestone aggregate for road building, as an ingredient of cement or as the starting material for the preparation of builder's lime by burning in a kiln. Walenta *et al.* [55] have successfully attempted the use of CA in cement compositions for designing civil engineering materials which can capture CO_2 from air and can sequester it as solid carbonates within the porous coatings of building walls. Calcium carbonate also finds its application in the purification of iron in a blast furnace and also in oil industry where it is added to drilling fluids as a formation-bridging and filtercake-sealing agent. It is also used as an extender in paints and commonly used as filler in plastics.

Mahinpey *et al.* [56] investigated the application of CA catalysed mineralization for sequestration of CO_2 in saline formations. He proposed the use of CA in minimizing the risk of leakage of CO_2 from the injection wellbore. Sufficient volume of enzyme solution can be injected at the end of CO_2 injection period through the injection wellbore. When the CO_2 would come in contact with the enzyme solution, solid particles would precipitate resulting in reduced permeability of porous media near the wellbore.

ENZYMATIC CAPTURE OF CARBON INTO METHANOL

Methanol is the simplest of alcohols and one of the oldest and most versatile of all renewable sources of energy. Thus, the conversion of atmospheric CO_2 to methanol offers a promising new technology for an efficient generation of fuel. This would be a carbon neutral approach which means that producing and burning this fuel will not increase CO_2 in the atmosphere. Methanol generation by partial hydrogenation of CO_2 has been accomplished by electrocatalysis and photocatalysis. A distinctive approach for the direct conversion of gaseous CO_2 to methanol involves the use of three dehydrogenases. The process involves initial reduction of CO_2 to formate catalysed by formate dehydrogenase (FateDH), followed by reduction of formate to formaldehyde by formaldehyde dehydrogenase (FaldDH), and finally formaldehyde is reduced to methanol by alcohol dehydrogenase (ADH).

The process is made thermodynamically favourable by using excess of nicotinamide adenine dinucleotide (NADH). Reduced NADH acts as a terminal electron donor for each dehydrogenase-catalysed reduction [57]. Obert and Dave [58] have successfully encapsulated these dehydrogenases in sol-gel matrix using the biocompatible synthesis method. The pores of the sol-gels act as nano-reactors for biocatalytic reactions. Although at present, this research is in preliminary stage, a detailed understanding of the reaction kinetics of the system remains to be fully understood.

CONCLUSION

As a result of global warming, the average temperature of the earth has increased by about 0.8°C. Almost 200 nations agreed in 2010 to circumvent hazardous effects of the climate change by limiting the rise of global earth's temperature to below 2°C (3.6 Fahrenheit). To achieve this goal, the world will have to slow down the rate of carbon emissions by an unprecedented rate or will have to sequester carbon emitted into the atmosphere. Most research is being focused on capturing CO_2 from the industries where it is emitted from, and also by collecting it directly from the air. Abiotic processes for carbon sequestration denotes capture, concentration, transport, and storage of CO_2 in deep oceans, underground geological structures, unmineable coal mines, and oil wells. These processes have large potential for storing carbon, but at the same time these are associated with huge costs and leakage risks.

Biotic processes provide natural, cost-effective, and secure ways to mitigate climate change risk. It has been proven that agricultural and forestry practices can partially mitigate increasing CO_2 concentration by sequestering carbon through photosynthesis. Microbes thriving at every possible niche on the earth have important roles to play in the global carbon and other mineral cycles. These are blessed with highly efficient enzymes, some of which are exceptionally good at capturing CO_2. Methanogens are able to utilize CO_2 and convert it into CH_4 and hydrogen gas. Carbonic anhydrase, a potent biocatalyst, provides a natural way of efficiently converting CO_2 to bicarbonate. Several companies are attempting to develop technologies by using the enzyme in a biomimetic approach for scrubbing CO_2 from industrial flue gases. The use of carbonic anhydrase as a biocatalyst in this process greatly reduces the costs normally associated with carbon capture. The present-day need is to understand and exploit the existing elegant biological processes of carbon capture combining biochemical engineering techniques for providing a natural and cost-effective process of carbon sequestration.

Summary

Biomimetic approach for carbon sequestration or CCS involves identification of a biological process or structure and its application to solve a non-biological problem. Carbonic anhydrases are the fastest enzymes known for their efficiency in converting CO_2 into bicarbonate. Efforts are underway for using carbonic anhydrases from various microbial sources for CO_2 sequestration. In this chapter, recent developments in immobilization of microbial carbonic anhydrases for their recycling and biological carbon sequestration are discussed. The possibility of an on-site scrubber that would provide a plant-by-plant solution to CO_2 sequestration, apart from eliminating the concentration and transportation costs, is the potential advantage of the biomimetic approach.

The photosynthetic fixation of atmospheric CO_2 in plants and trees could be of great value in maintaining a CO_2 balance in the atmosphere. Algal systems, on the other hand, being more efficient in photosynthetic capabilities, are the choice of research for solving the global warming problem. The biomass thus produced could be used for various purposes. Non-photosynthetic CO_2 fixation occurs widely in nature by the methanogenic archaea. A mixture of three cultures of bacteria (Rhodospirillum rubrum, Methanobacterium formidium *and* Methanosarcina barkeri) *has been attempted for complete bioconversion of oxides of carbon to methane. Dual benefit of carbon sequestration along with the production of useful amino acids such as lysine employing microbes* (Corynebacterium glutamicum) *makes this approach very attractive.*

REFERENCES

1. IPCC (Intergovernmental Panel on Climate Change). 2001. *Climate Change: Mitigation.* Cambridge: Cambridge University Press

2. Puri, A. K. and T. Satyanarayana. 2010. Carbon sequestration for mitigating disastrous effects of global warming. In *Natural and Man Made Disasters*, K. K. Singh and A. K. Singh (eds), pp. 229–252. New Delhi: MD Publications

3. Puri, A. K. and T. Satyanarayana. 2010. Enzyme and microbe mediated carbon sequestration. *CO_2 Sequestration Technologies for Clean Energy* S. Z. Qasimand M. Goel (eds), pp.119–129. Delhi: Daya Publishing House

4. Kindermann, G., M. Obersteiner, B. Sohngen, J. Sathaye, K. Andrasko, E. Rametsteiner, B. Schlamadinger, S. Sven Wunder, and R. Robert Beach. 2008. Global cost estimates of reducing carbon emissions through avoided deforestation. *Proceedings of the National Academy of Sciences of the United States of America* 105:10302–7

5. Christer, J., S. D. Wullschleger, U. C. Kalluri, and G. A. Tuskan. 2010. Phytosequestration: Carbon biosequestration by plants and the prospects of genetic engineering. *Bio Science* 60:685–696

6. Dahlman, R. C., G. K. Jacobs, and F. B. Metting Jr. 2001. What is the potential for carbon sequestration by the terrestrial biosphere? *First National Conference on Carbon Sequestration* 14–17 May, Washington, DC

7. Parr, J. F. and L. A. Sullivan. 2005. Soil carbon sequestration in phytoliths. *Soil Biology and Biochemistry* 37:117–24

8. Min, D. H. 2011. Carbon sequestration potential of switchgrass as a bio-energy crop. Michigan State University. Details available at. http://msue. anr. msu.edu/news/ carbon_ sequestration_potential_of_ switchgrass_ as_a_bio_energy_crop_

9. Kubler, J. A., A. M. Johnston, and J. A. Raven. 1999. The effects of reduced and elevated CO_2 and O_2 on the seaweed *Lomentaria articulata*. *Plant Cell Environ.* 22:1303–1310

10. Yue, L. and W. Chen. 2005. Isolation and determination of cultural characteristics of a new highly CO_2 tolerant fresh water microalgae. *Energy Conversion and Management* 46:1868–1876

11. Wood, H. G., C. H. Werkman, A. Hemingway, and A. O. Nier. 1941. Heavy carbon as a tracer in heterotrophic carbon dioxide assimilation. *J. Biol. Chem.* 139:375–367

12. Norici, A., A. Dalsass, and M. Giordano. 2002. Role of phospho-enol-pyruvate carboxylase in anaplerosis in the green microalga Dunaliella salina cultured under different nitrogen regimes. *Physiol. Plant.* 116:186–191

13. Sauer, U. and B. J. Eikmanns. 2005. The PEP-pyruvate-oxaloacetate node as the switch point for carbon flux distribution in bacteria. *FEMS Microbiol. Rev.* 29:765–94

14. Gubler, M., S. M. Park, M. Jetten, G. Stephanopoulos, and A. J. Sinskey. 1994. Effects of phosphoenol pyruvate carboxylase deficiency on metabolism and lysine production in *Corynebacterium glutamicum*. *Applied Microbiol. and Biotechnol.* 40:857–863

15. Lee, J. C., J. H. Kim, W. S. Chang, and D. Pak. 2012. Biological conversion of CO_2 to CH_4 using hydrogenotrophic methanogen in a fixed bed reactor. *J. Chem. Technol. Biotechnol.* 87:844–847

16. Khalifah, R. G. and D. N. Silverman. 1991. *The Carbonic Anhydrases: Cellular Physiology and Molecular Genetics*, S. J. Dodgson, R. E. Tashian, G. Gros, and N. D. Carter (eds), pp. 49–70. New York: Plenum Press

17. CO_2 Solution Inc. 2012. Enzymatic power for carbon capture. *The Process*. Details available at http://www.co2solutions.com/en/the-process

18. Bhattacharya, S., M. Schiavone, S. Chakrabarti, and S. K. Bhattacharya. 2003. CO_2 hydration by immobilized carbonic anhydrase. *Biotechnology and Applied Biochemistry* 38:111–117

19. Iliuta, I. and F. Larachi. 2012. New scrubber concept for catalytic CO_2 hydration by immobilized carbonic anhydrase II & in-situ inhibitor removal in three-phase monolith slurry reactor. *Separation and Purification Technol.* 86:199–214

20. Larachi, F., O. Lacroix, and B. P. A. Grandjean. 2012. Carbon dioxide hydration by mobilized carbonic anhydrase in Robinson-Mahoney and packed-bed scrubbers. *Chemical Engineering Science* 73:99–115

21. Ge. J., R. M. Cowan, C. Tu, M. L. McGregor, and M. C. Trachtenberg. 2002. Enzyme-based CO_2 capture for advanced life support. *Life Support & Biosphere Science* 8:181–189

22. Cowan, R. M., J. J. Ge, Y. J. Qin, M. L. McGregor, and M. C. Trachtenberg. 2003. CO_2 capture by means of an enzyme-based reactor. *Ann. New York Acad. Sci.* 984:453–469

23. Matsuyama, H., M. Teramoto, and K. Iwai. 1994. Development of a new functional cation-exchange membrane and its application to facilitated transport of CO_2. *Journal of Membrane Science* 93:237–244

24. Matsuyama, H., A. Terada, T. Nakagawara, Y. Kitamura, and M. Teramoto. 1999. Facilitated transport of CO_2 through polyethylenimine/poly (vinyl alcohol) blend membrane. *Journal of Membrane Science* 163:221–227

25. Trachtenberg, M. C., M. Cowan, and D. A. Smith. 2009. Membrane-based, enzyme-facilitated, efficient carbon dioxide capture. *Energy Procedia* 1:353–360

26. Trachtenberg, M. C., D. A. Smith, R. M. Cowan, and X. Wang. 2007. Flue gas CO_2 capture by means of a biomimetic facilitated transport membrane. *Proceedings of the AIChE Spring Annual Meeting*, Houston, Texas, USA

27. Bond, G. M., J. Stringer, D. K. Brandvold, F. A. Simsek, M. G. Medina, and G. Egeland. 2001. Development of integrated system for biomimetic CO_2 sequestration using the enzyme carbonic anhydrase. *Energy and Fuels* 15:309–316

28. Alber, B. E. and J. G. Ferry. 1994. Carbonic anhydrase from the archaeon *Methanosarcina thermophila*. *Proceedings of the National Academy of Sciences of the United States of America* 91:6909–6913

29. Smith, K. S. and J. G. Ferry. 1999. A plant-type (b-class) carbonic anhydrase in the thermophilic methanoarchaeon *Methanobacterium thermoautotrophicum*. *J Bacteriol* 181:6247–6253

30. Borchert, M. and P . Saunders. 2011. Heat stable carbonic anhydrases and their use. *U.S. Patent US7892814*

31. Newman, L. M., L. Clark, C. Ching, and S. Zimmerman. 2010. Carbonic anhydrase polypeptides and uses thereof. *US Patent WO10081007*

32. Daigle, R. and M. Desrochers. 2009. Carbonic anhydrase having increased stability under high temperature conditions. *US Patent 7521217*

33. Prabhu, C., S. Wanjari, S. Gawande, S. Das, S. Labhsetwar, S. Kotwal, A. K. Puri, and T. Satyanarayana. 2009. Immobilization of carbonic anhydrase enriched microorganism on biopolymer based materials. *Journal of Molecular Catalysis B.* 60:13–21

34. R. Yadav, T. Satyanarayana, N. Labhsetwar, S. Rayalu. 2011. Immobilization of carbonic anhydrase on chitosan beads for enhanced carbonation reaction. *Process Biochemistry* 46:1010–1018

35. Prabhu, C., A. Valechha, and S. Wanjari. 2011. Carbon composite beads for immobilization of carbonic anhydrase. *Journal of Molecular Catalysis B.* 71:71–78

36. Sharma, A., A. Bhattacharya, and A. Shrivastava. 2011. Biomimetic CO_2 sequestration using purified carbonic anhydrase from indigenous bacterial strains immobilized on biopolymeric materials. *Enzyme and Microbial Technology* 48:416–426

37. Prabhu, C., S. Wanjari, A. Puri, A. Bhattacharya, R. Pujari, R. Yadav, S. Das, N. Labhsetwar, A. Sharma, T. Satyanarayana, and S. Rayalu. 2011. Region-specific bacterial carbonic anhydrase for biomimetic sequestration of carbon dioxide. *Energy and Fuels* 25:1327–1332

38. Wanjari, S., C. Prabhu, T. Satyanarayana, A. Vinu, and S. Rayalu. 2012. Immobilization of carbonic anhydrase on mesoporous aluminosilicate for carbonation reaction. *Microporous Mesoporous Mater* 160: 151–158

39. Yadav, R. R., S. N. Mudliar, A. Y. Shekh, A. B. Fulke, S. S. Devi, K. Krishnamurthi, A. Juwarkar, and T. Chakrabarti. 2012. Immobilization of carbonic anhydrase in alginate and its influence on transformation of CO_2 to calcite. *Process Biochemistry* 47:585–590

40. Yadav, R., S. Wanjari, and C. Prabhu. 2010. Immobilized carbonic anhydrase for the biomimetic carbonation reaction. *Energy and Fuels* 24:6198–6207

41. Yadav, R., T. Satyanarayana, S. Kotwal, and S. Rayalu. 2011. Enhanced carbonation reaction using chitosan-based carbonic anhydrase nanoparticles. *Current Science* 100:520–524

42. Kanbar, B. and E. Ozdemir. 2010. Thermal stability of carbonic anhydrase immobilized within polyurethane foam. *Biotechnology Progress* 26:1474–1480

43. Sharma, A. and A. Bhattacharya. 2010. Enhanced biomimetic sequestration of CO_2 into $CaCO_3$ using purified carbonic anhydrase from indigenous bacterial strains. *Journal of Molecular Catalysis B.* 67:122–128

44. Premkumar, L., U. K. Bhageshwar, I. Irena Gokhman, A. Zamir, and J. L. Sussmanb. 2003. An unusual halotolerant a-type carbonic anhydrase from the alga *Dunaliella salina* function-ally expressed in *Escherichia coli. Protein Expression and Purification* 28:151–157

45. Kaur, S., M. N. Mishra, and A. K. Tripathi. 2010. Gene encoding gamma-carbonic anhydrase is cotranscribed with argC and induced in response to stationary phase and high CO_2 in *Azospirillum brasilense* Sp7. *BMC Microbiology* 10:184

46. Kim, I. G., B. H. Jo, D. G. Kang, C. S. Kim, Y. S. Choi, and H. J. Cha. 2012. Biomineralization-based conversion of carbon dioxide to calcium carbonate using recombinant carbonic anhydrase. *Chemosphere* 87:1091–1096

47. Lee, R. B. Y., J. A. C. Smith, and R. E. M. Rickaby. 2013. Cloning, expression and characterization of the δ-carbonic anhydrase of *Thalassiosira weissflogii* (Bacilli-riophyceae). *J. Phycol.* 49:170–177

48. Chi, Z., J. V. O'Fallon, and S. Chen. 2011. Bicarbonate produced from carbon capture for algae culture. *Trends Biotechnol* 29:537–41

49. Farrell, A. 2011. Carbon dioxide storage in stable carbonate minerals. Basalt laboratory studies of interest to carbon capture and storage. Advisor MN Evans University of Maryland Geology p 1–24

50. Mirjafari, P., K. Asghari, and N. Mahinpey. 2007. Investigation the application of enzyme carbonic anhydrase for CO_2 sequestration purposes. *Ind. Eng. Chem. Res.* 46:921–926

51. Gutknecht, J., M. A. Bisson, and F. C. Tosteson. 1977. Diffusion of carbon dioxide through lipid bilayer membrane. *J. Gen. Physiol.* 69: 779–794

52. Favre, N., M. L. Christ, and A. C. Pierre. 2009. Biocatalytic capture of CO_2 with carbonic anhydrase and its transformation to solid carbonate. *J. Mol. Catal. B: Enzymatic* 60:163–170

53. Bond, G. M., G. Egeland, D. K. Brandvold, M. G. Medina, F. A. Simsek, and J. Stringer. 1999. Enzymatic catalysis and CO_2 sequestration. *World Res. Rev.* 11:603–619

54. Soong, Y., D. L. Fauth, and B. H. Howard. 2006. CO_2 sequestration with brine solution and fly ashes. *Energy Conversion and Management* 47:13–4, 1676–1685

55. Walenta, G., V. Morin, A. C. Pierre, and L. Christ. 2011. Cementitious compositions containing enzymes for trapping CO_2 into carbonates and/or bicarbonates. PCT International Application. WO 2011048335 A1 2011042

56. Mahinpey, N., K. Asghari, and K. Mirjafari. 2011. Biarbonate produced from carbon capture for algae culture. *Chemical Engineering Research and Design* 89:1873–1878

57. Dave, B. C. 2008. Prospects in methanol production. In *Bioenergy*, J. Wall, C. S. Harwood, and A. L. Demain (eds), pp. 235–245. Washington D. C.: ASM Press

58. Obert, R. and B. C. Dave. 1999. Enzymatic conversion of carbon dioxide to methanol: Enhanced methanol production in silica sol–gel matrices. *Journal of the American Chemical Society* 121:12192–193

WEBSITES

1. www.energy.ca.gov/glossary/
2. http://www.carbozyme.us/pub abstracts.html

CHAPTER 12

Bio-sequestration of CO_2 – Potential and Challenges

K. Uma Devi[1], G. Swapna[1] and K. Suman[1,2]
[1]Department of Botany, Andhra University, Visakhapatnam-530003
[2]Centre for Marine Living Resources and Ecology, Kendriya Bhavan,
PB No. 5415, Kochi-682037
E-mail: umadevikoduru@gmail.com

INTRODUCTION

The disastrous consequences arising from increasing levels of anthropogenic carbon dioxide (CO_2) emissions and the predicted catastrophe, if unabated, are now widely recognized. The 2007 Nobel Peace Prize was shared by the Intergovernmental Committee on Climate Change (IPCC) and Al Gore for their "efforts to build-up and disseminate greater knowledge about man-made climate change and to lay the foundations for the measures that are needed to counteract such change". The documentary film *An Inconvenient Truth* is about Al Gore's campaign to educate citizens about global warning on climate change. The documentary won two Academy Awards in 2007. After the ratification of Kyoto Protocol by 191 states in 2005, research has geared up to explore and develop appropriate methods for CO_2 capture and storage (long-term/permanent)–dubbed as CCS (carbon capture and storage/sequestration). CCS has evolved to CCUS or CCSU, with the insertion of 'utilization' in this programme aiming to utilize the captured carbon for commercial purposes.

Fossil fuel-fired power plants are responsible for 40% of global CO_2 emissions [1]. The flue gases contain, besides CO_2, other greenhouse gases (GHGs) such as oxides of nitrogen (NO_x) and oxides of sulphur (SO_x) which cause acid rain. Developing an optimal system to capture and subsequently store or utilize CO_2 in the flue gases from the power plants is crucial in managing CO_2 levels in the atmosphere. Carbon sequestration is being attempted through physical and chemical means.

Of late, biological sequestration is also being considered. Each of these approaches has a potential, but there are several technical and fiscal challenges as well. Technical issues relate to mode of transfer of flue gases to the algal culture medium and means of harvest of microalgae. Fiscal concerns are with regard to the cost of set up of the facility and the running costs.

Biological sequestration involves the use of living organisms – plants, because they use CO_2 for the synthesis of carbohydrates in the process called photosynthesis. Some of the methods used are discussed in the following sections.

Terrestrial Sequestration

Afforestation of barren lands can serve as CO_2 sinks. Growing more trees and replacing felled trees with new saplings (reforestation) is an age old tradition. In realization of the impending high global temperatures in the near future, research efforts are about to recognize the plant species, which would survive and flourish under high temperatures. At higher temperatures, the microscopic pores (stomata) on leaves through which CO_2 diffuses into the plant get bunged to avoid evaporation of water (through a process called transpiration). Plant species tolerant to high CO_2 and temperatures are, therefore, being identified through simulated experimental studies of CO_2 and environment interactions on plant growth [2].

Ocean Fertilization

The ocean is considered to be the largest sink for CO_2. The microscopic algae (phytoplankton) which constitute the primary producers in the marine ecosystem, utilize CO_2 for photosynthesis. About 6%–8% of the atmospheric carbon is believed to be fixed by them [3]. Iron is a mineral present in limiting amounts to support phytoplankton growth. Fertilization of ocean with iron, termed as ferrigation (hypothesis proposed by Martin [4]), is believed to promote luxurious growth of phytoplankton and, thereby, increase the sequestration of CO_2. Smetacek *et al.* [5] reported that ferrigation induced diatom-dominated phytoplankton blooms accompanied by considerable CO_2 fixation in the ocean surface layer. The fate of bloom biomass could not be resolved in these experiments. This is because of the mass mortality of the diatom species in the bloom and it was assumed that at least half of the bloom biomass sank to a depth of 1000 m. It was proposed that iron-fertilized diatom blooms might fix carbon for centuries at the bottom of the ocean. However, this idea casted several apprehensions on its efficacy, the ratio of iron added to carbon sequestered and various side effects of ocean ferrigation. Moreover, the idea is unpopular with the public, because it is perceived as meddling with nature [6].

MASS CULTURE OF MICROALGAE

Cultivation of microalgae on a mass-scale is also perceived as a measure of large-scale bio-sequestration of CO$_2$. This would be a good option for the CCSU programme as the algal biomass can be used for various purposes. The opportunities and challenges of this method are herewith discussed.

Microalgae are small (microscopic) photosynthetic microorganisms. They constitute a large group in the living world, represented by thousands of species. They can be unicellular, filamentous, or colonial (Figure 1). Microalgae are aquatic and live in all kinds of water – fresh, sea, estuarine, and sewage water; species that thrive in different waters

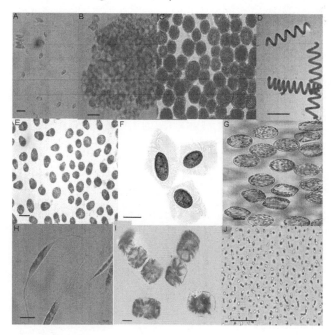

Figure 1 *Images of microalgal species: (A)* Scenedesmus dimorphus, *(B)* Desmodesmus sp., *(C)* Haematococcus pluvialis, *(D)* Arthrospira platensis, *(E)* Dunaliella salina, *(F)* Tetraselmis *sp., (G)* Odontella aurita, *(H)* Cylindrotheca fusiformis, *(I)* Thalassiosira *sp. and (J)* Synechococcus *sp. Bar represents 10 μm.*

being different. They have the following features that make them good candidates for carbon sequestration:

(i) They multiply very fast with a generation time of less than a day under suitable environmental conditions.

(ii) The rate of photosynthesis in these organisms is much higher than the land plants. They can thus fix CO_2 more efficiently compared to land plants.

(iii) The CO_2 enters the cells passively through diffusion and there is no regulated entry through stomata as in land plants. Hence, CO_2 uptake remains uninterrupted even at high temperatures compared to land plants in which stomata closes at high temperatures.

(iv) Algal biomass can be used for various economically important products (Table 1). One of the perceived potential that has been realized is in nutraceuticals – nutrient with positive therapeutic and health benefits. There are species rich in proteins, pigments like β carotene, lutein, astaxanthin and lipids with omega fatty acids (polyunsaturated fatty acids-PUFAs) [7–8]. Of late, there is a renewed interest in exploring microalgae as a source of biofuel [9–11]. The concept of biofuel from microalgae has drawn the attention of many due to its various advantages [12].

The algal biomass can be used for various biotechnological purposes as listed in Table 1.

CO_2 MITIGATION FROM FLUE GAS USING MICROALGAE

Flue gas is a mixture of CO_2, NO_X (70–420 ppm), and SO_X (50–400 ppm). Microalgae can assimilate all the constituents of the flue gas mixture, because they can utilize NO_X as their nitrogen supplement for growth, CO_2 for photosynthesis, and they can tolerate and absorb up to 300 ppm SO_X. The NO_X dissolves in water and becomes available in an assimilatory form to the microalgae. Many industrial set-ups and thermal power plants are installed with sulphur scrubbers as a mandatory pollution check practice. Land plants can utilize only CO_2, but not the NO_X component of flue gases.

A proof of concept experiment on bio-sequestration of CO_2 in flue gases was conducted by Isaac Berzin at MIT Boston, USA funded by the Green Fuel Technologies [30]. The flue gas from the chimneys of the thermal power plant near the MIT campus was fed to a microalgal culture system set-ups on the terrace of an MIT building (Figure 2).

Table 1 : Survey of work on biotechnological uses of microalgal biomass

Microalgal species	Biotechnological products
Spirulina platensis	Protein and pigments [13–14]
Chlorella protothecoides	Nutraceutical (Lutein) and biofuel [15]
Dunaliella salina	Nutraceutical (β-carotene) and mariculture [16–17]
Haematococcus pluvialis	Nutraceutical (Astaxanthin) [18]
Skeletonema costatum	Mariculture [19]
Pavlova lutheri	Mariculture [19]
Isochrysis galbana	PUFAs and mariculture [19]
Porphyridium sp.	PUFAs and polysaccharide [20]
Nannochloropsis sp.	PUFAs and mariculture [21]
Tetraselmis suecica	Mariculture, biofuel [19]
Chaetoceros calcitrans	Mariculture [19]
Phaeodactylum tricornutum	Mariculture [19] and biofuel [22]
Crypthecodinium cohnii	PUFAs [23]
Schizochytrium sp.	Mariculture [19]
Synechoccus sp.	Bioactive compounds [24]
Botryococcus braunii	Biofuel [25]
Chlamydomonas rheinhardii	Biohydrogen and biofuel [26]
Neochloris oleoabundans	Biofuel [27]
Nannochloropsis sp.	Biofuel [28]
Euglena gracilis	Biotin [29]
Pleurochyrsis cartarae	Biofuel [19]

Note: Number in brackets denote reference.

An international network on biofixation of CO$_2$ for GHG abatement using microalgae was proposed in 2001 for the development of this technology [31]. The feasibility of using microalgae bioreactors at an industrial scale to sequester CO$_2$ from power plant exhaust gases (flue gas) has been considered [32].

Following the proof of concept demonstration of Berzin, several large-scale set-ups have been established. Electric company (Israel Electric Corporation–IEC) in Ashkelon, Israel utilizes sulphur dioxide-free flue gas for the cultivation of algae. It has a continuous supply of free, filtered, and chlorinated sea water that can be obtained at the rate of 450,000 m^3/hr.

Figure 2   *A photobioreactor with microalgal culture fed by flue gas, from the chimney of thermal power plant on the terrace of a building at MIT campus, Boston*
Source: http://news.cnet.com/8301-11128_3-10239916-54.html

Seambiotic Ltd has designed systems for the removal of high SO_X content in the flue gas mixture from coal-fired power plant in IEC, by using flue gas desulphurization (FGD) techniques. FGD treated flue gas is supplied to salt water algae raceway ponds for the growth of *Nannochloropsis* sp. The algae showed increased growth rate with FGD treated flue gas than pure CO_2 (US patent number US2008/0220486 A1).

US patent 5659977 by Cyanotech Corporation, Hawaii has described methods of using CO_2 from exhaust gas of a fossil fuel-fired power plant for algae production. 188 kg/hr CO_2 from the flue gas chimney is transferred to the bottom of a CO_2 absorption tower (6.4 m high packing material), which can provide 67 tonnes of CO_2 per month that aids in the production of 36 tonnes per month of Spirulina. The patent also describes the utilization of heat from fossil fuel engine to dry the algal biomass and the electrical energy is used to drive motors, pumps and also to provide illumination for algal growth.

Nature Beta Technologies Ltd, Eilat City, Israel reported *Dunaliell asalina* biomass production of 20 $g/m^2/d$. The cost of dry biomass, when cultured with a flue gas mixture was estimated at USD 0.34/kg as compared to USD 17/kg cultivation in normal environment [33].

Rheinisch-Westfälisches Elektrizitätswerk's algae project, Germany erected at the Niederaussem power plant location, utilized desulphurized flue gases for algal growth, which was operated for three years until 2011 (Source: http://www.rwe.com/).

Microalgal Cultivation with Flue Gases: Opportunities

Microalgal cultivation poses no competition to agriculture both in terms of usage of water and land. Freshwater is not required for many species, sea water can be used to culture marine and estuarine species. Even sewage water can be used, when the algae are cultivated for biofuel [34]. Cultivation facility – a shallow, circular or a raceway pond – can be set up on barren lands, even deserts; fertile land is not required [35–36].

Sequestration of carbon from flue gas emissions through physical and chemical methods requires separation of CO_2 from the other gases. A major component of flue gases is NO_X, which is also a GHG, but is not captured in physical and chemical methods of carbon sequestration. In bio-sequestration with microalgae, both the CO_2 and the NO_X in the flue gas are utilized – the former for photosynthesis and the latter as a source of nitrogen, a nutrient required for growth. The NO_X dissolves in water and is converted into an assimilatory form to the microalgae. Jiang *et al.* [37] reported that *Scenedesmus dimorphus* has a tolerance to high concentration of CO_2 (20%), NO_X (150–500 ppm), and SO_X (100 ppm). *Nannochloris* sp. is reported to grow under 100 ppm of nitric oxide (NO) [38]. *Tetraselmis* sp. was found to flourish, when supplied with flue gas with 185 ppm of SO_X and 125 ppm of NO_X in addition to 14.1% CO_2 [39]. *Dunaliella tertiolecta* was found to grow well in flue gas with 1000 ppm of NO and 15% CO_2 concentration assimilating 51%–96% of NO depending on the growth condition [40]. Maeda *et al.* [41] examined the tolerance of a strain of *Chlorella* and found that the strain could grow in the presence of different combinations of trace elements.

Flue gas cooling, compression, transport, and supply to the algal mass cultivation units like open raceway ponds or photobioreactor constitute a major share of the total production cost. It is not always possible to set up an algal cultivation system adjacent to flue gas chimneys in the thermal power plant units or in the industries, because of space constraint. A high level of dust and other air pollutants in the industrial units also have an impact on the algae cultivated in the open ponds. Moreover, photobioreactors do not have an effective control on emission of CO_2

from exhaust gases, because CO_2 is usually bubbled through the reactor with the excess CO_2 being emitted to the atmosphere and this technology is very expensive to operate [42]. An alternate strategy is to enrich water with flue gases until saturation and transport the water to algal cultivation units located away from industrial units. This method avoids cooling and compression costs of flue gases, and algal production units need not be constructed near power plants. If freshwater is used, the pH of the water falls to as low as 2 (our unpublished results), because of dissolution of CO_2 (carbonic acid), NO_X, and SO_X. For sea water (pH 8) even after continuous supply of flue gas for two continuous days, the pH does not change much (7–7.5) perhaps due to the buffering capacity of sea water [43]. The pH of flue gas enriched water can be adjusted to optimum, based on the type of algae that is cultivated. The general optimum pH for freshwater algae is 6 and for marine algae, it is around 8.

There are reports of experiments of testing growth of microalgae in flue gas enriched water [44–47]. In the laboratory at Andhra University, the growth rate of some freshwater and marine microalgae in water enriched with flue gas emissions from a gas-fired furnace of the steel (Visakhapatnam Steel Plant, Visakhapatnam) were studied. The results were encouraging (Table 2).

Microalgal Cultivation with Flue Gases: Challenges

In most cases, the flue gases have to be transported to the microalgal cultivation site away from the industrial set-up. This is true for physical and chemical methods of carbon sequestration from flue gases. The major cost in carbon sequestration involves these transportation costs.

Microalgae can be cultivated in closed or open systems. The closed systems are called photobioreactors and they are designed in various configurations like flat-plate and tubular reactors (Figure 3). They are highly automated with controlled temperature, pH, and other physical and chemical conditions. Therefore, there is optimal growth of algae and high biomass productivity. Also, since it is a contained system, there is no problem of contamination of the algal biomass by other algal species or bacteria. Photobioreactors are being used for the cultivation of *Haematococcus pluvialis* [48], *Tetraselmis suecica* [40], *Nannochloropsis* sp. [9], and *Chlorella vulgaris* [50].

The drawback with photobioreactors is scalability, with maximum unit size of about 100 m^2 and the cost of construction. The costs become

Table 2 : Growth of different microalgal species within nutrient medium made from flue gas enriched water (fresh/sea) in comparison to normal water (fresh/sea) and CO_2 enriched water (fresh/sea)

Microalgal species	% Increase (+)/Decrease (–) in growth in flue gas enriched water compared to		Remarks
	Normal water	CO_2 enriched water	
*Chlorella protot**hecoides*	+50	+19	Flue gas enriched water is more effective than pure CO_2
Scenedesmus dimorphus	+193	+21	Flue gas enriched water is more effective than pure CO_2 alone
Desmodesmus sp.	+36	NA	NA
Haematococcus pluvialis	+25	-16	Pure CO_2 supply is more effective than flue gases
Neochloris oleoabundans	+38	+35	Flue gas enriched water has profound influence on growth than pure CO_2 supply
Dunaliella salina	0	+20	No advantage in flue gas enriched water*
Tetraselmis sp.	0	+18	No advantage in flue gas enriched water*

*Flue gas might not have dissolved in sea water as the pH did not fall from ~8 to not below 7.5.

Figure 3 *Photobioreactors: Designed in various configurations like flat-plate and tubular reactors*
Source: http://www.et.byu.edu/~wanderto/homealgaeproject/ Photobioreactor;http://www.uanews.org/story/biofuels-algae-hold-potential-not-ready-prime-time;http://www.algaeindustrymagazine.com/ aim-interview-asus-dr-milton-sommerfeld/

forbidding, far exceeding the value of the algal biomass. Cleaning and maintenance of photobioreactors is also very difficult with the algae often clinging to the walls of the reactor. Running a photobioreactor is an energy intensive process and nullifies the aim of carbon sequestration. Construction of photobioreactors to cater to sequestration of large volumes of flue gas is highly unrealistic both in terms of maintenance and in terms of economy.

Open pond systems are, therefore, the only option for mass cultivation of microalgae. These are shallow ponds, which are either lined with concrete or other materials or just left with the retaining mud walls (Figure 4). They are 10–50 cm deep which will allow good penetration of light. Natural sunshine is utilized. These ponds are fitted with paddle wheels for gas/liquid mixing and circulation. Such open pond systems are especially suitable in areas with abundant sunshine for most part of the year. Much less energy is utilized in operation (for paddle wheels) of raceway ponds. However, contamination by other microalgal species and microbes or grazers is an inevitable contingency. Consistent production levels are not possible, because of the variation in the climate such as temperature, rainfall and, so on. Therefore, hardy microalgal species can only be mass cultivated in open pond systems.

Figure 4   *Different open ponds for mass culture of microalgae: raceway and circular with paddle wheels*
Source: http://www.nature.com; http://www.nature.com nature/journal/v474/n7352_ supp/full/474S015a.html; http://www.et.byu. edu/~wanderto/homealgaeproject/ Photobioreactor.html

Though many microalgal species have potential for various biotechnologically useful products, commercial success has only been possible with three species – *Chlorella* (Centre Pivot Ponds, Taiwan and Japan), *Dunaliella salina* (Cognis-Hutt Lagoon, Western Australia; Nature Beta-Eilat, Israel) and *Haematococcus pluvialis* (Cyanotech-Kona, Hawaii; Algatech Ltd, Israel). These systems are typically used in commercial large-scale cultivation of algae – *Dunaliella salina* [51], *Arthrospira platensis* [52], *Pleurochysis carterae* [53], and *Nannochloropsis* sp. [54].

To commence mass culture in open ponds, seed culture is required. Seed culture is developed in the laboratory under controlled conditions. The culture is initiated in a small volume, which is gradually upscaled to larger volumes and the process takes several days (15–25) depending upon the scale of mass culture. Microalgae being photoautotrophic, grow to a level where dense cultures like bacteria and yeast are attained is not possible, and light becomes a limiting factor. Indoor culture of seed culture to feed outdoor ponds is an energy intensive process as it requires artificial light, temperature control, and so on. Moreover, flue gas mitigation using microalgae requires set-up of large-scale culture units.

To reduce the time and thus energy costs in generating seed culture, indoors mixotrophy culture can be resorted too. Mixotrophy involves culture under light in an inorganic nutrient medium supplemented with an organic carbon source [55]. The algae in such cultures photosynthesize and simultaneously use the ready carbon source in the medium. Because they are not exclusively dependent on light, dense growth occurs in mixotrophy cultures. The duration of the seed culture development time is substantially reduced compared to photoautotropical cultures. Glycerol and acetate are the most commonly used carbon sources, because they can be obtained as by-products from various industries [56–57]. Several microalgal species – *Chlorella vulgaris* [58], *Botryococcus braunii* [59], and *Phaeodactylum tricornutum* [22] have been reported to be adaptable to mixotrophic culture. *Scenedesmus dimorphus, Neochloris oleoabundans, Dunaliella salina, Chlorella protothecoides,* and *Desmodesmus* sp. were found to respond favourably to mixotrophic cultures in our studies (unpublished results). Thus, the unique feature of microalgae – mixotrophy can be used to develop seed culture in the laboratory to inoculate outdoor ponds, where CO_2/flue gases can be used for algal growth.

Harvesting algal biomass is yet another challenging task. It contributes to 20%–30% of total production costs [60]. The microalgal cultures are dilute (usually <0.5 kg/m³ dry mass) and large volumes of water have to be processed for harvesting the cells. Flocculation is followed by centrifugation; filtration or sedimentation techniques are also in practice. Filtration is not a practical approach for large volumes of cultures and head loss of culture is more in this method. Centrifugation is comparatively costly and can only be used for high value products [61]. Flocculation technique is widely used in several mass culture systems. Multivalent metal salts such as alum have been widely used to flocculate algal biomass in wastewater treatment processes [62–65]. Alum is an effective flocculent for *Scenedesmus* and *Chlorella* [66]; however, flocculation by metal salts may be unacceptable if biomass is to be used in certain aquaculture and other applications. Edible, non-toxic polymeric flocculants like chitosan are reported to be effective with *Tetraselmis chui, Isochrysis* sp., and *Thalassiosira pseudonana* [67]. An optimized method of harvesting is yet to be designed for each algal species. More research is required in the area of flocculants.

The production costs of algal biomass are high even when cultivated in open pond systems. That is why economic viability is possible only for high value products like those used as nutraceuticals (beta carotene with a global market of USD 247 million, astaxanthin USD 200 million, lutein USD 233 million, and PUFAs USD 700 million). As of now, the economic feasibility of microalgae as source of biofuel looks bleak [68–70]. The scientist Maverick Craig Venter is engaged in research on microalgal biofuel [71]. There is thus a beacon of hope that a technology breakthrough may result in the realization of flue gas sequestration to produce carbon neutral renewable source of fuel. A considerable amount of production costs can perhaps be compensated via carbon credits, which was formalized in the Kyoto Protocol.

CONCLUSION

Bio-sequestration of GHGs in power plant flue gas through microalgal mass culture has been conceptualized and is being experimented, because of several advantages. However, it has enormous challenges to surmount. It can be said that containment of GHGs in the modern industrial world is imperative to abate further dire consequences of climate change. A plurality of approaches is required for this task and bio-sequestration through microalgae thus remains a potential option.

Summary

Methods of managing the release of CO_2 through sequestration by physical and chemical methods have been worked out. As techniques of CO_2 containment are prohibitively expensive, the prospect of biological carbon fixation for remediation of CO_2 is being explored. Plants are known as CO_2 sinks – they fix CO_2 during daytime. Tiny (microscopic) aquatic plants, the microalgae, have even more efficient CO_2 fixing ability than large plants and trees. They have a very quick growth rate. Some of these species are rich in oil suitable for converting to diesel (fuel) and with nutrient properties (with omega fatty acids). Some species have carotenes and other pigments; some are rich in proteins. They can be mass cultured in lands, not suitable for agriculture and their biomass can be beneficially used in production of biodiesel, nutracueticals for humans, and as animal feed. Flue gas (a mixture of CO_2 and the other GHG NO_X) released from industries can be used to ferrigate the mass culture units of microalgae thus effectively sequestering them. The proofs of concept experiments of this bio-sequestration option have been demonstrated in 2005. Large-scale industries based on this principle are yet to take-off, despite the lure of biodiesel. The challenges in this technology are many, including both technical and fiscal.

REFERENCES

1. Carapellucci, R. and A. Milazzo. 2003. Membrane systems for CO_2 capture and their integration with gas turbine plants. *Proceedings of the Institution of Mechanical Engineers, Part A: Journal of Power and Energy* 217:505–517

2. Poorter, H. and M. Perez-Soba. 2002. Plant growth at elevated CO_2. *The Earth System: Biological and Ecological Dimensions of Global Environmental Change* 2:489–496

3. Reeburgh, W. S. 1997. Figures summarizing the global cycles of biogeochemically important elements. *Bulletin of the Ecological Society of America* 260–267

4. Martin, J., K. Coale, K. Johnson, S. Fitzwater, and R. Gordon 1994. Testing the iron hypothesis in ecosystems of the equatorial Pacific Ocean. *Nature* 371:123–129

5. Smetacek, V., C. Klaas , V. H. Strass, P. Assmy, and M. Montresor. 2012. Deep carbon export from a Southern Ocean iron-fertilized diatom bloom. *Nature* 487:313–319

6. Smetacek, V. and S. Naqvi. 2008. The next generation of iron fertilization experiments in the Southern Ocean. *Philosophical Transactions of the Royal Society A: Mathematical, Physical and Engineering Sciences* 366:3947–3967

7. Dunstan, G. A., J. K. Volkman, S. M. Barrett, J. M. Leroi, and S. Jeffrey. 1993. Essential polyunsaturated fatty acids from 14 species of diatom (Bacillariophyceae). *Phytochemistry* 35(1):155–161

8. Spolaore, P., C. Joannis-Cassan, E. Duran, and A. Isambert. 2006. Commercial applications of microalgae. *Journal of Bioscience and Bioengineering* 101:87–96

9. Chiu, S. Y., C. Y. Kao., M. T. Sai, S. C. Ong, and C. H. Chen. 2009. Lipid accumulation and CO_2 utilization of *Nannochloropsis oculata* in response to CO_2 aeration. *Bioresource Technology* 100:833–838

10. Kumar, A., S. Ergas, X. Yuan, A. Sahu, and Q. Zhang. 2010. Enhanced CO_2 fixation and biofuel production via microalgae: Recent developments and future directions. *Trends in Biotechnology* 28:371–380

11. Colin, V. L., A. Rodríguez, and H. A. Cristóbal. 2011. The role of synthetic biology in the design of microbial cell factories for biofuel production. *BioMed Research International* (published online)

12. Schenk, P. M., S. R. Thomas-Hall, E. Stephens, U. C. Marx, and J. H. Mussgnug. 2008. Second generation biofuels: High-efficiency microalgae for biodiesel production. *Bioenergy Research* 1:20–43

13. Qiang, H. 2004. Industrial production of microalgal cell-mass and secondary products-major industrial species. *Handbook of Microalgal Culture: Biotechnology and Applied Phycology*, p. 264. Chichester, UK: Wiley

14. Prasanna, R., A. Sood, P. Jaiswal, S. Nayak, and V. Gupta. 2010. Rediscovering cyanobacteria as valuable sources of bioactive compounds (Review). *Applied Biochemistry and Microbiology* 46:119–134

15. Iwamoto, H. 2004. Industrial production of microalgal cell-mass and secondary products-major industrial species. *Handbook of Microalgal Culture: Biotechnology and Applied Phycology* A. Richmond (ed) p. 255. Chichester, UK: Wiley

16. Ben-Amotz, A. 2004. Industrial production of microalgal cell-mass and secondary products-major industrial species. *Handbook of Microalgal Culture: Biotechnology and Applied Phycology*, p. 273. Chichester, UK: Wiley

17. Cysewski, G. and R. T. Lorenz. 2004. Industrial production of microalgal cell-mass and secondary products-species of high potential. *Haematococcus. Handbook of Microalgal Culture: Biotechnology and Applied Phycology*, p. 281. Chichester, UK: Wiley

18. Becker, W. 2004. Microalgae for aquaculture–The nutritional value of microalgae for aquaculture. *Handbook of Microalgal Culture: Biotechnology and Applied Phycology* A. Richmond (ed). Chichester, UK: Wiley

19. Arad, S. and A. Richmond. 2004. Industrial production of microalgal cell-mass and secondary products-species of high potential. *Handbook of Microalgal Culture: Biotechnology and Applied Phycology* p. 289. Chichester, UK: Wiley

20. Zittelli, G. C., L. Rodolfi, and M. R. Tredici. 2004. Industrial production of microalgal cell-mass and secondary products-species of high potential. *Handbook of Microalgal Culture: Biotechnology and Applied Phycology* p–298

21. Liu, X., S. Duan , A. Li, N. Xu, Z. Cai Z, and Z. Hu. 2009. Effects of organic carbon sources on growth, photosynthesis and respiration of *Phaeodactylum tricornutum. Journal of Applied Phycology* 21:239–246

22. Mendes, A., A. Reis, R. Vasconcelos, P. Guerra, and T. L. da Silva. 2009. Crypthecodinium cohnii with emphasis on DHA production: Areview. *Journal of Applied Phycology* 21(2):199–214

23. Abed, R., S. Dobretsov, and K. Sudesh. 2008. Applications of cyanobacteria in biotechnology. *Journal of Applied Microbiology* 106:1–12

24. Yoo, C., S. Y. Jun, J. Y. Lee, C. Y. Ahn, and H. M. Oh. 2010. Selection of microalgae for lipid production under high levels carbon dioxide. *Bioresource Technology* 101:S71–S74

25. Beer, L. L., E. S. Boyd, J. W. Peters, and M. C. Posewitz. 2009. Engineering algae for biohydrogen and biofuel production. *Current Opinion in Biotechnology* 20(3):264–271

26. Levine, R. B., M. S. Costanza-Robinson, and G. A. Spatafora. 2011. *Neochloris oleoabundans* grown on anaerobically digested dairy manure for concomitant nutrient removal and biodiesel feedstock production. *Biomass and Bioenergy* 35(1):40–49

27. Gouveia, L. and A. C. Oliveira. 2009. Microalgae as a raw material for biofuels production. *Journal of Industrial Microbiology & Biotechnology* 36:269–274

28. Baker, E. R, J. J. McLaughlin, S. H. Hutner, B. De Angelis, and S. Feingold. 1981. Water-soluble vitamins in cells and spent culture supernatants of *Poteriochromonas stipitata, Euglena gracilis* and *Tetrahymena thermophila. Archives of Microbiology* 129:310–313

29. Brown, L. M. 1996. Uptake of carbon dioxide from flue gas by microalgae. *Energy Conversion and Management* 37:1363–1367

30. http://news.cnet.com/8301-11128_3-10239916-54.html

31. Pedroni, P., J. Davison, H. Beckert, P. Bergman, and J. Benemann. 2001. A proposal to establish an international network on biofixation of CO$_2$

and greenhouse gas abatement with microalgae. *Journal of Energy and Environmental Research* 1:136–150

32. Nakamura, T., C. Senior, M. Olaizola, and S. Masutani. 2001. Capture and sequestration of CO_2 from stationary combustion systems by photosynthesis of microalgae, Proc. First National Conf. on Carbon Sequestration, Department of Energy–National Energy Technology Laboratory, May 2001

33. Benemann, J., B. Koopman, J. Weissman, D. Eisenberg, and R. Goebel. 1980. Development of microalgae harvesting and high rate pond technology. *Algal Biomass*. G. S. C. Shelef (ed), pp. 457–499. Amsterdam: Elsevier

34. Chinnasamy, S., A. Bhatnagar, R. W. Hunt, and K. Das. 2010. Microalgae cultivation in a wastewater dominated by carpet mill effluents for biofuel applications. *Bioresource Technology* 101:3097–3105

35. Bell, R. A. 1993. Cryptoendolithic algae of hot semiarid lands and deserts. *Journal of Phycology* 29:133–139

36. Demirbas, M. F. 2011. Biofuels from algae for sustainable development. *Applied Energy* 88(10):3473–3480.

37. Jiang, Y., W. Zhang, J. Wang, Y. Chen, S. Shen, and T. Liu 2013. Utilization of simulated flue gas for cultivation of *Scenedesmus dimorphus*. *Bioresource Technology* 128:359–364

38. Yoshihara, K. I., H. Nagase, K. Eguchi, K. Hirata, and K. Miyamoto. 1996. Biological elimination of nitric oxide and carbon dioxide from flue gas by marine microalgae NOA-113 cultivated in a long tubular photobioreactor. *Journal of Fermentation and Bioengineering* 82:351–354

39. Matsumoto, H., N. Shioji, A. Hamasaki, Y. Ikuta, and Y. Fukuda. 1995. Carbon dioxide fixation by microalgae photosynthesis using actual flue gas discharged from a boiler. *Applied Biochemistry and Biotechnology* 51:681–692

40. Nagase, H., K. Eguchi, K. I. Yoshihara, K. Hirata, and K. Miyamoto. 1998. Improvement of microalgal NO_X removal in bubble column and airlift reactors. *Journal of Fermentation and Bioengineering* 86:421–423

41. Maeda, K., M. Owada, N. Kimura, K. Omata, and I. Karube. 1995. CO_2 fixation from the flue gas on coal-fired thermal power plant by microalgae. *Energy Conversion and Management* 36:717–720

42. Jeong, M. L., J. M. Gillis, and J. Y. Hwang. 2003. Carbon dioxide mitigation by microalgal photosynthesis. *Bulletin-Korean Chemical Society* 24:1763–1766

43. Ben-Yaakov, S. 1973. pH buffering of pore water of recent anoxic marine sediments. *Limnology and Oceanography* p. 86–94

44. Negoro, M., A. Hamasaki, Y. Ikuta, T. Makita, and K. Hirayama. 1993. Carbon dioxide fixation by microalgae photosynthesis using actual flue gas discharged from a boiler. *Applied Biochemistry and Biotechnology* 39:643–653

45. Gao, K., Y. Aruga, K. Asada, and M. Kiyohara. 1993. Influence of enhanced CO_2 on growth and photosynthesis of the red algae *Gracilaria sp.* and *G. chilensis*. *Journal of Applied Phycology* 5:563–571

46. Hu, H. and K. Gao. 2003. Optimization of growth and fatty acid composition of a unicellular marine picoplankton, *Nannochloropsis* sp. with enriched carbon sources. *Biotechnology Letters* 25(5):421–425

47. Yue, L. and W. Chen. 2005. Isolation and determination of cultural characteristics of a new highly CO_2 tolerant fresh water microalgae. *Energy Conversion and Management* 46:1868–1876

48. García-Malea, M., F. Acién, J. Fernández, M. Cerón M, and E. Molina. 2006. Continuous production of green cells of *Haematococcus pluvialis:* Modelling of the irradiance effect. *Enzyme and Microbial Technology* 38:981–989

49. Chini Zittelli, G., L. Rodolfi, N. Biondi, and M. R. Tredici. 2006. Productivity and photosynthetic efficiency of outdoor cultures of *Tetraselmis suecica*in annular columns. *Aquaculture* 261:932–943

50. Ugwu, C., J. Ogbonna, and H. Tanaka. 2002. Improvement of mass transfer characteristics and productivities of inclined tubular photobioreactors by installation of internal static mixers. *Applied Microbiology and Biotechnology* 58:600–607

51. García-González, M., J. Moreno, J. C. Manzano, F. J. Florencio, and M. G. Guerrero. 2005. Production of *Dunaliella salina* biomass rich in 9 cis-β-carotene and lutein in a closed tubular photobioreactor. *Journal of Biotechnology* 115:81–90

52. Richmond, A. 1992. Mass culture of cyanobacteria. *Photosynthetic prokaryotes*, N. Mann and N. Carr (eds), pp. 181–210. New York and London: Plenum Press

53. Moheimani, N. R. and M. A. Borowitzka. 2006. The long-term culture of the coccolithophore Pleurochrysis carterae (Haptophyta) in outdoor raceway ponds. *Journal of Applied Phycology* 18:703–712

54. Cheng-Wu, Z., O. Zmora, R. Kopel, and A. Richmond. 2001. An industrial-size flat plate glass reactor for mass production of *Nannochloropsis* sp.(*Eustigmatophyceae*). *Aquaculture* 195:35–49

55. Ogawa, T. and S. Aiba. 1981. Bioenergetic analysis of mixotrophic growth in *Chlorella vulgaris* and *Scenedesmus acutus. Biotechnology and Bioengineering* 23:1121–1132

56. Garcıa-Ochoa, F., V. Santos, J. Casas, and E. Gomez. 2000. Xanthan gum: Production, recovery and properties. *Biotechnology Advances* 18:549–579

57. Chen, C. Y., K. L. Yeh, R. Aisyah, D. J. Lee, and J. S. Chang. 2011. Cultivation, photobioreactor design and harvesting of microalgae for biodiesel production: A critical review. *Bioresource Technology* 102(1):71–81

58. Heredia-Arroyo, T., W. Wei, R. Ruan, and B. Hu. 2011. Mixotrophic cultivation of *Chlorella vulgaris* and its potential application for the oil accumulation from non-sugar materials. *Biomass and Bioenergy* 35:2245–2253

59. Zhang, H., W. Wang, Y. Li, W. Yang, and G. Shen. 2011. Mixotrophic cultivation of *Botryococcus braunii*. *Biomass and Bioenergy* 35:1710–1715

60. Gudin, C. and C. Thepenier. 1986. Bioconversion of solar energy into organic chemicals by microalgae. *Advances in Biotechnological Processes* p. 6

61. Molina Grima, E., E. H. Belarbi, F. Acien Fernandez, A. Robles Medina, and Y. Chisti. 2003. Recovery of microalgal biomass and metabolites: Process options and economics. *Biotechnology Advances* pp. 20:491–515

62. McGarry, M. G. 1970. Algal flocculation with aluminum sulphate and polyelectrolytes. *Journal (Water Pollution Control Federation)* pp. 191–201

63. Dodd, J. C. 1979. Algae production and harvesting from animal wastewaters. *Agricultural Wastes* 1:23–37

64. Benemann, J. R. 1997. CO_2 mitigation with microalgae systems. *Energy Conversion and Management* 38:S475–S479

65. Moraine, R., G. Shelef, E. Sandbank, Z. Bar-Moshe, and G. Shvartzburd. 1980. Recovery of sewage-borne algae: Flocculation, flotation and centrifugation techniques. *Algae Biomass,* G. S. C. Shelef (eds), pp. 531–545. Amsterdam: Elsevier/North-Holland Biomedical Press

66. Golueke, C. G. and W. J. Oswald. 1965. Harvesting and processing sewage-grown planktonic algae. *Journal (Water Pollution Control Federation)* 471–498

67. Heasman, M., J. Diemar , W. O'connor, T. Sushames, and L. Foulkes. 2000. Development of extended shelf-life microalgae concentrate diets harvested by centrifugation for bivalve Molluscs–A summary. *Aquaculture Research* 31:637–659

68. Milledge, J. J. 2011. Commercial application of microalgae other than as biofuels: A brief review. *Reviews in Environmental Science and Biotechnology* 10:31–41

69. Stephens, E., I. L. Ross, Z. King, J. H. Mussgnug, and O. Kruse. 2010. An economic and technical evaluation of microalgal biofuels. *Nature Biotechnology* 28:126–128

70. Ribeiro, L. A. and P. P. da Silva. 2012. Technoeconomic assessment on innovative biofuel technologies: The case of microalgae. *ISRN Renewable Energy* Vol 2012, Article ID 173753, 8 pages. doi: 10.5402/2012/173753

71. http://www.forbes.com/sites/christopherhelman/2012/06/24/milking-oil-from-algae-craig-venter-makes-progress-in-exxon-backed-venture/

CHAPTER 13

Cyanobacteria in Carbon Dioxide Utilization and as Biosurfactants and Flocculants

Romi Khangembam, Minerva Shamjetshabam, Gunapati Oinam, Angom Thadoi Devi, Aribam Subhalaxmi Sharma, Pukhrambam Premi Devi, and O.N. Tiwari

Microbial Bioprospecting Laboratory, Microbial Resources Division, Institute of Bioresources and Sustainable Development
(An Autonomous Institute under the DBT, Government of India)
Takyelpat, Imphal-795001, Manipur, India
E-mail: ontiwari1968@gmail.com

INTRODUCTION

Cyanobacteria are considered as the most primitive photosynthetic prokaryotes, which are supposed to have appeared on this planet during the Pre-Cambrian period [1]. They are referred to in the literature by various names, chief among which are Cyanophyta, Myxophyta, Cyanochloronta, Cyanobacteria, blue-green algae, and blue-green bacteria. They are the oldest known fossils and account for almost 8% of the total population of bacteria in water bodies. This is significant because 3500 million years ago, the earth was a little over 1000 million years old itself with conditions that were vastly different from those of the modern day. Fossils of 1000 million years of age are met with some scepticism and the actuality of these fossils being cyanobacteria has been disputed [2].

Cyanobacteria are a unique assemblage of organisms, which occupy and predominate a vast array of habitats as a result of several general characteristics; some belonging to bacteria and others unique to higher plants [3–4]. Cyanobacteria are susceptible to sudden physical and chemical alterations of light, salinity, temperature, and nutrient composition [5–6]. They are a group of Gram-negative photoautotrophic microorganisms capable of oxygenic photosynthesis [7]. These ancient organisms [8] have minimal nutrient requirements. They are capable of fixing carbon dioxide (CO_2) and utilizing light as an energy source and water as an electron donor [9]. In addition, many strains possess the ability to fix atmospheric nitrogen [7, 9]. Such minimal requirements enable cyanobacteria to inhabit nearly every illuminated environment on earth, aquatic or terrestrial, from the Arctic deserts to hydrothermal hot springs, to oceans, to freshwater environments [7, 10–11]. Cyanobacteria are able to compensate for changes in the earth's atmosphere and local environments by investing in a carbon concentrating mechanism (CCM) to adjust to changes in CO_2 concentrations and some are able to form heterocyst for changes in nitrogen availability.

CYANOBACTERIA IN CARBON CAPTURE AND UTILIZATION

Cyanobacterial progenitors first appeared some 2.7 billion years ago, when the oxygen level in the atmosphere was low and CO_2 levels were manifold greater than today [12]. However, it is near certain that cyanobacteria have been subjected to periods of rapid evolutionary change throughout the past 2.7 billion years. In particular, the marked drop in CO_2 levels and rise in oxygen levels that occurred around 400–350 million years ago [12–13], combined with the diffusional resistance to CO_2 transfer in liquid, may have triggered adaptations to cope with photorespiration and low-efficiency carbon gain. These adaptations would have included transport mechanisms for active uptake of inorganic carbon (Ci) and subsequent localized elevation of CO_2 around the primary carboxylating enzyme, ribulose bisphosphate carboxylase-oxygenase (Rubisco), and the partitioning of Rubisco into icosahedral, micro-compartments known as carboxysomes [14–16]. Collectively, these CCM adaptations function to raise the concentration of CO_2 around Rubisco and, thereby, improve the efficiency of CO_2 fixation, suppressing the wasteful oxygenase reaction of Rubisco.

In response to ancient changes in atmospheric CO_2 and oxygen levels, the cyanobacteria (blue-green algae) evolved an environmental adaptation, known as a CCM, which has a significant positive effect on photosynthetic performance. The CCM functions to actively transport inorganic carbon species (Ci, HCO_3^-, and CO_2) resulting in conditional accumulation of bicarbonate (HCO_3^-) within the cell. This pool of HCO_3^- is thereafter utilized to provide elevated CO_2 concentrations around the primary CO_2 fixing enzyme, Rubisco, which is in turn encapsulated in unique micro-compartments known as carboxysomes. Cyanobacteria are found in many niches, which arguably have applied evolutionary pressure on the precise compositional diversity of cyanobacterial CCMs.

During the past two centuries, the atmospheric concentration of greenhouse gases or GHGs (for example CO_2, methane) increased significantly [17]. As a result, the global temperature is rising and it is projected that a total CO_2 reduction of 50%–85% is required by 2050 in order to stabilize the emission within the "safe zone" of 450 ppm [18]. Bio-based solutions to this problem have been implemented with success, but also with challenges. First-generation biofuel (that is ethanol) derived from sugar cane and corn has been commercially viable; however, its negative impacts on food supply have raised ethical concerns about its long-term consequences. Improvement of carbon fixation has been a subject of long and extensive research over the past four decades. Oxygenic phototrophs, including cyanobacteria, algae, and plants, use the reductive pentose phosphate cycle, also known as the Calvin-Benson-Bassham cycle (hereafter Calvin cycle), for assimilation of CO_2 (Figure 1). Numerous efforts to improve the rate of CO_2 fixation via the Calvin cycle, PEP carboxylase, and through synthetic pathway introduction have been attempted with varying degrees of success. The approaches can be largely grouped into the following four categories:

(i) Engineering of ribulose-1, 5-bisphosphate carboxylase/oxygenase (RuBisCO) for improvements of the rates of catalysis of carboxylation and reduction of the oxygenation reaction.

(ii) Enhancement of the activation state of RuBisCO.

(iii)Improvements of the regeneration phase of the Calvin cycle.

(iv)Enrichment of CO_2 concentration around RuBisCO for suppression of the oxygenase reaction.

In the cyanobacteria, mechanisms exist that allow photosynthetic CO_2 reduction to proceed efficiently even at very low levels of Ci. These inducible, active transport mechanisms enable the cyanobacteria to accumulate large internal concentrations of inorganic carbon that may be up to 1000-fold higher than the external concentration. As a result, the external concentration of Ci required to saturate cyanobacterial photosynthesis in vivo is orders of magnitude lower than that required to saturate the principal enzyme (ribulose bisphosphate carboxylase) involved

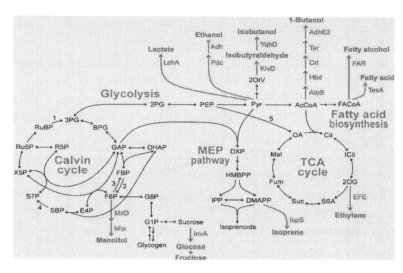

Figure 1 *Central metabolic pathways of cyanobacteria. Heterologous pathways and products that have been genetically engineered in cyanobacteria and discussed in the text are indicated in magenta.* Abbreviations: Enzymes: 1, ribulose-1, 5 bisphosphate carboxylase/oxygenase (RuBisCO); 2, fructose-1, 6-bisphosphatase; 3,phosphofructokinase;4,sedoheptulose-1,7-bisphosphatase;5, phosphoenolpyruvate carboxylase. Metabolites: 2OG, 2-oxoglutarate; 2OIV, 2-oxoisovalerate; 2PG, 2 phosphoglycerate; 3PG, 3-phosphoglycerate; AcCoA, acetyl-CoA; BPG, 1, 3-bisphosphoglycerate; Cit, citrate; DHAP, dihydroxyacetone-phosphate; DMAPP, dimethylallyl-pyrophosphate; DXP, 1-deoxyxylulose-5-phosphate; E4P, erythrose-4-phosphate; F6P, fructose-6-phosphate; FACoA, fatty acyl-CoA; FBP, fructose-1, 6-bisphosphate; Fum, fumarate; G1P, glucose-1-phosphate; G6P, glucose-6-phosphate; GAP, glyceraldehyde-3-phosphate; HMBPP, 1-hydroxy-2-methyl-2-butenyl-4-pyrophosphate; ICit, isocitrate; IPP, isopentenyl-pyrophosphate; Mal, malate; OA, oxaloacetate; PEP, phosphoenolpyruvate; Pyr, pyruvate; R5P, ribose-5-phosphate; Ru5P, ribulose-5-phosphate; RuBP, ribulose- 1, 5-bisphosphate; S7P, sedoheptulose-7-phosphate; SBP, sedoheptulose-1, 7-bisphosphate; SSA, succinic semialdehyde; Suc, succinate; X5P, xylulose-5-phosphate.

in the fixation reactions. Since CO_2 is the substrate for carbon fixation, the cyanobacteria somehow perform the neat trick of concentrating this small, membrane permeable molecule at the site of CO_2 fixation.

Over the last few years, climate change has been identified as one of the biggest problems facing mankind, prompting policy makers to act in response to both scientific and social movements. One of the consequences of these movements is the development of the Kyoto Protocol, which makes signatory nations obligatory to take actions to reduce GHG emissions [19]. CO_2 emissions related to world energy demand or fossil fuel combustion (coal, petroleum, or natural gas) are increasing by 2.1% per year [20] and although different strategies have been proposed to reduce this, no single solution will be sufficient to solve it. A reduction in power consumption or improvements to combustion processes can help reduce CO_2 emissions. Alternatively, carbon capture and storage (CCS) shows great potential in diminishing the amount of CO_2 released into the atmosphere from combustion processes [21–22]. Other than these, replacing fossil fuels with renewable energy sources can reduce CO_2 emissions by up to 80% (International Energy Outlook). Thus, great effort has been focused on the production of biodiesel, bioethanol, or biogas. The use of biofuels does not contribute to an increase in the atmospheric CO_2 concentration since the carbon released on combustion has previously been taken from the atmosphere by photosynthesis. However, if CO_2 from biofuel combustion was captured and stored, a net removal of CO_2 from the atmosphere would result [23].

Biological systems could potentially make a significant contribution to CCS technology, as they can be deployed in a sustainable, renewable manner. Photosynthetic microbes are an attractive option for biological CCS because they have the ability to capture sunlight and use that energy to store carbon in forms useful to humans such as fuels, food additives, and medicines. The use of algae-like cyanobacteria, cultured in non-oceanic environments, to capture emissions directly from fossil fuel sources could be a technology that aids in the inexpensive reduction of CO_2 emissions from the energy sector over the coming decades. The ultimate carbon emissions associated with deployment of such a technology would depend on the capacity and efficiency of the algae to capture the carbon and on the use of the stored carbon after capture. Algae capture CO_2 and fix it into carbon molecules using photosynthetic processes similar to land plants (Figure 2). Photosynthesis uses the energy from light to reduce carbon from CO_2 to complex carbon-based molecules that act as stored energy. These molecules are often fuels, fuel precursors, or high value chemicals.

The capacity of plants to contribute to biofuel production is much lower than that required; thus, other alternatives such as microalgae, like cyanobacteria, have been proposed [24–25]. They can grow faster than plants, do not require fertile land or useful water, reach solar energy utilization efficiencies up to 5%, and are able to use direct flue gases as their carbon source. The production of 1 tonne of biomass requires a minimum consumption of 1.8 tonnes of CO_2, so CO_2 can

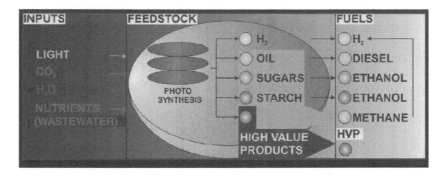

Figure 2 *Overview of inputs and outputs of photosynthesis in algae.*
Note: Light, CO_2 and water are utilized by the photosynthetic reactions to produce valuable products such as hydrogen (H_2), oil (including triacylglycerols; TAG's), sugars and starch. H_2 can be utilized as a fuel or light energy carrier in fuel cells. Oils, sugars and starches can be converted into fuels such as diesel, alcohols or methane. High value products (HVP) can be made to supplement costs and include products such as carotenoids, antibodies or commodities such as organic acids

be captured from flue gases and transformed into biomass, valuable biofuel products such as carotenoids, amino acids, biofuels, and so on, or used in bioremediation processes [26–29]. The utilization of cultures as a CO_2 biofixation method was proposed more than 50 years ago [30] and it has been commonly accepted that blue-green algae can contribute to CO_2 capture and storage. Then, how can microalgae such as cyanobacteria contribute to a reduction in CO_2 emissions? The answer is by (a) producing biofuels to replace the fossil fuels used today or alternatively and (b) allowing the production of other commodities or by-products from flue gases, which allow one to obtain credits to enhance the economic yield of CCS. There are two further ways in which microalgae might contribute in reducing the present-day CO_2 emissions both related to wastewater and residuals treatment, that is (c) a reduction in power consumption due to the aeration in integrated wastewater treatment processes by using a bacteria–microalgae consortium, and (d) enhancement in the quality of biogas produced in these plants by reducing their CO_2 content using cultures.

Microalgae which include large cyanobacteria have several advantages over crops in the production of biofuels as they do not compete for fertile land or water, thus do not affect food supply and other crop products. In this sense, they have been proposed as the only alternative for the sustainable production of biodiesel and bioethanol [25]. However, to be competitive in the bioenergy market, the production of cultures must approximate crop price and capacity. Palm oil, for instance, is produced at a volume of 40 Mt per year and has value of 0.5 Euro/kg (€/kg) in the global market. It is important to note that approximately 40% of the dry weight of biomass is carbon so the production of 100 tonnes of biomass requires a minimum of 183 tonnes of CO_2; but because the biomass produced cannot be stored for a long time, it is not a sequestration strategy. The decomposition of biomass releases the previously fixed carbon into the atmosphere resulting in this biomass also being considered as an energy vector.

Genetic modification of CCMs may improve the energetic efficiency and rate of carbon uptake in oxygenic photosynthetic organisms. Green algae and cyanobacteria have evolved mechanisms to uptake and concentrate Ci from the environment. The strategy utilized depends on the form of carbon encountered. Conversion of CO_2 to HCO_3^- in an aqueous environment is pH dependent, with basic environments promoting formation of HCO_3^-. Within the cell, enzymatic inter-conversion takes place in order to transport and concentrate CO_2 at the place of carbon fixation in the chloroplast pyrenoid in green algae or carboxysome in cyanobacteria.

CYANOBACTERIAL EXTRACELLULAR POLYMERIC SUBSTANCES

Cyanobacteria can be found in almost any ecological niche from freshwater and salt water to terrestrial and extreme environments, including metal-contaminated habitats [31–32]. Most of the strains are able to produce extracellular polymeric substances (EPS) mainly of polysaccharidic nature. The EPS can remain associated to the cell surface as sheaths, capsules and/or slimes or be liberated into the surrounding environment as released polysaccharides (RPS) [33]. The EPS serve as a boundary between the cyanobacterial cell and its immediate environment. Polysaccharides are characterized by an extreme structural diversity; as a result, they play very diverse roles in nature and may get modified under stress conditions [34]. Numerous bacterial polysaccharides are potentially available, known to be involved in pathogenesis, symbiosis, biofilm formation, protection from phagocytic predation and stress resistance in microorganisms; but relatively few have been commercially established [35–36]. The presence of proteins, uronic acids, pyruvic acid, and *O*-methyl, *O*-acetyl and sulphate groups emphasizes the complex nature of cyanobacterial EPS. Cyanobacteria, which are the most widespread photosynthetic bacteria, produce large amounts of EPS [37–38]. The cyanobacterial EPS can be divided in two main groups – the one associated with the cell surface and the other as polysaccharides released into the surrounding environment as RPS.

The available data on the monosaccharidic composition of cyanobacterial EPS reveal some peculiar features of these polymers, when compared with those produced by other microorganisms, such as the presence of one or two uronic acids, constituents rarely found in the EPS produced by other microbial groups. Cyanobacterial EPS also contain sulphate groups, a feature unique among bacteria, but shared by the EPS produced by archaea and eukaryotes. Both the sulphate groups and the uronic acids contribute to the anionic nature of the EPS, conferring a negative charge and a 'sticky' behaviour to the overall macromolecule [39–40]. The anionic charge is an important characteristic for the affinity of these EPS towards cations, notably metal ions. However, the ability to chelate metal ions is related not only to the amount of charge groups, but also to their distribution on the macromolecules and their accessibility [40–41]. On the other hand, many cyanobacterial EPS are also characterized by a significant level of hydrophobicity, which is due to the presence of ester-linked acetyl groups (up to 12% of EPS dry weight), peptidic moieties, and deoxysugars such as fucose and rhamnose.

Cyanobacteria are also used in aquaculture, wastewater treatment, food, fertilizers, production of secondary metabolites including exopolysaccharides, vitamins, toxins, enzymes, and pharmaceuticals [42]. The monosaccharide most frequently found at the highest concentration in cyanobacterial EPS is glucose, although there are polymers where other sugars, such as xylose, arabinose, galactose or fucose, are present at higher concentrations than glucose [43]. Indeed, the interest in cyanobacteria as producers of high-molecular weight polysaccharides is related to the capability of these biopolymers to modify the rheological properties of water, acting as thickening agents and to stabilize the flow properties of aqueous solutions. According to the chemical and physicochemical features of the cyanobacterial exopolysaccharides, there are possible fields of application in pharmaceutical industry, because of their antiviral or immuno-stimulating properties or the capability of slowly releasing drugs [44]. However, despite their ubiquitous distribution, very little data regarding the physicochemical properties of cyanobacterial EPS exists.

CYANOBACTERIAL EPS AS BIOFLOCCULANTS AND BIOSURFACTANTS

Flocculants have been defined as agents, which are used in fast solid–liquid separations. Their addition enables dispersed particles to aggregate together and form flocs of a size to make them settle speedily and to clear the system. They are said to act on a molecular level on the surfaces of the particles to reduce repulsive forces and increase attractive forces and mounting attractive forces.

Amongst bioremediants, bioflocculants have gained increasing attention since they are environmentally friendly, biodegradable, and non-toxic. Bioflocculants contain various organic groups, such as uronic acids (containing a carbonyl and a carboxylic acid function), glutamic and aspartic acid in the protein component or galacturonic acid and glucuronic acid in the polysaccharide component, which are responsible for binding metals. The use of bioflocculants in wastewater treatment seems to be an economical alternative to physical and chemical means. Unlike mono-functional ion-exchange resins and other physicochemical methods, bioflocculants are capable of removing inorganic/organic particles through their flocculating activity. *Phormidium* sp. Strain J-I, a benthic filamentous cyanobacterium, has been found to produce significant amounts of extracellular flocculants. This flocculant is a sulphated heteropolysaccharide, in which fatty acids and protein are bound. Extracellular flocculants were also found to be produced by the benthic cyanobacteria *Anabaena circularis* PCC 6720. Two planktonic cyanobacteria, *Anabaena* sp. N1444 and *Anabaena* sp. PC-1 can also produce extracellular flocculants.

On the basis of renewable biomass they are safe and biodegradable, cheap and non-toxic, but show weak activity in applications. Since the benthic photoautotrophic cyanobacteria occupy a low-light zone, water-clarifying bioflocculants are important to the photosynthetic activity of these organisms. Because of their agglutinating properties, the bioflocculants may also remove soluble nutrients from the water column, resulting in heterotrophic activity in the sediment region. Cyanobacteria possess a unique cell wall that combines the presence of an outer membrane and lipopolysaccharides, as in Gram-negative bacteria, with a thick and highly cross-linked peptidoglycan layer similar to Gram-positive bacteria. However, a lack of information regarding both the genes encoding the proteins involved in the EPS biosynthetic pathways and the factors controlling these processes strongly limits their potential for biotechnological applications.

Surfactants are compounds that reduce the surface tension of liquid, the interfacial tension between two liquids, or that between a liquid and a solid. Biosurfactants encompass the properties of dropping surface tension, stabilizing emulsions, promoting foaming, and are usually non-toxic and biodegradable. These molecules have a potential to be used in a variety of industries such as cosmetics, pharmaceuticals, food preservatives, and detergents. Presently, the production of biosurfactants is highly expensive due to the use of synthetic culture media. Biosurfactants or surface-active compounds are a heterogeneous group of surface active molecules produced by microorganisms, which either adhere to cell surface or are excreted extracellulary in the growth medium. Chemically synthesized surfactants are not biodegradable and can be toxic to environment. Biosurfactants have special advantage over their commercially manufactured counterparts because of their lower toxicity, biodegradable nature, and effectiveness at extreme temperature, pH, salinity, and ease of synthesis. They are potential candidate for much commercial application in the pharmaceutical, food processing, and oil recovery industries. Other potential applications of biosurfactants relate to food, cosmetic, health care industries, and cleaning toxic chemicals of industrial and agricultural origin.

Biosurfactants are used highly as biosorbents. Metal sequestering by different parts of the cell can occur via various processes – complexation, chelation, coordination, ion exchange, precipitation, and reduction. Microorganisms possess an abundance of functional groups that can passively adsorb metal ions. The removal of toxic hexavalent chromium from aqueous solution by biosorption by different biomass types has been extensively reported. Higher metal adsorptive capacity of EPS of *Lyngbya putealis* as compared to its immobilized biomass suggests it to be a better biosorbent for metal removal [45]. Also soft metals such as gold and palladium are first bound on sites and within the cell wall, and these sites act as nucleation points for the reduction of metals and growth of crystals and elemental gold and palladium have been obtained. The conventional methods to remove these toxic heavy metals are generally expensive and not satisfactorily effective at trace levels of heavy metal contamination. During the past decades, a promising new method using microbes as a biosorbent for heavy metal removal has emerged and received considerable attention.

Cyanobacteria thus represent one of the potential choices for biological treatment of wastewater because they increase oxygen content of water via photosynthesis and remove heavy metals via adsorption/absorption processes. Bala *et al.* demonstrated that many functional groups (such as carboxyl, hydroxyl, sulphate, and other charged groups) making up the cyanobacterial cell wall played important roles in metal removal. These functional groups are parts of carbohydrates, proteins, and lipids of cyanobacterial mucilage that cover their cell surface. The major part of mucilage is exocellular EPS found both in culture medium and on cell surface (capsular polysaccharides or CPS). They act as the binding sites for metals and the capacity of cyanobacteria in removing metal ions depends on their affinity and specificity to metal ions. Various techniques have been developed to remove lead from the environment. These include the applications of chemical and biological materials such as cyanobacteria. Many kinds of cyanobacteria were used to remove heavy metals in Thailand. The cells of this cyanobacterium are covered with a mucilage sheath. Thus, we speculated that it would have high capacity for removing metals from solution and serve as one of the potential alternative biomaterials for removal of metals from contaminated water.

CONCLUSION

Currently, CCS technologies are energy intensive and expensive, often requiring specific operating conditions of temperature, pressure, pH, and concentration in addition to specific citing, where the properties of the sub-surface allow CO_2 storage. Worldwide, there are several carbon storage demonstrations, pilot and proposed projects underway to better understand the processes and challenges to carbon storage technologies [46]. Cyanobacteria could be an alternative or additional technology to this, if the overall running costs are less and implementation at scale can be achieved. However, considerations such as land availability and water use may limit the potential for the sustainable deployment of them as a means of capturing carbon from fossil fuel sources globally. The ability to couple cyanobacterial growth to power plant water use and to use low-grade heat from the power plant, technological advances that bring harvesting and extraction costs down and ultimately finding or engineering organisms that show an order of magnitude increase in efficiency will all impact the extent of deployment of this technology and its successful competition with other carbon capture technologies. Also, fundamental studies on cyanobacterial polysaccharides are relatively few in number compared to other polysaccharides. The data currently available indicate cyanobacterial exopolysaccharides have considerable biotechnological importance, but further studies are necessary to evaluate the feasibility of their practical application.

A major challenge in future commercial exploitation of cyanobacteria is the product yield. Design and operational practices of cultivations and harvests of the products hold the keys to economic feasibility, as do strain optimization for higher productivities. Substantial improvements in carbon fixing and recycling pathways are needed for the cyanobacterial cell factory to become a commercial reality and to offer an economically and environmentally sustainable production system.

Summary

Cyanobacteria are the simplest form of algae representing plant kingdom. They are photosynthetic prokaryotic organisms, which are unicells or filaments. Among these, certain cyanobacteria are able to compensate for changes in the earth's atmosphere and local environments by investing in a CCM and contribute to CO_2 capture and utilization. Most of the strains are also able to produce extracellular EPS mainly of polysaccharidic nature. According to the chemical and physicochemical features of the cyanobacterial exopolysaccharides, there are possible biotechnological fields of application for these polymers as biosurfactants and flocculants.

REFERENCES

1. Ash, N. and M. Jenkins. 2006. Biodiversity and poverty reduction: The importance of biodiversity for ecosystem services. Final report prepared by the United Nations Environment Programme World Conservation Monitoring Centre (UNEP-WCMC) (http://www.unepwcmc.org) for the Department for International Development (DFID)

2. Brasier, M. D., O. R. Green, A. P. Jephcoat, A. K. Kleppe, M. J. Van Kranendonk, J. F. Lindsay, A. Steele, and N. V. Grassineau. 2002. Questioning the evidence for Earth's oldest fossils. *Nature* 416:76–81

3. Abd Allah, L. S. 2006. Metal-binding ability of cyanobacteria: The responsible genes and optimal applications in bioremediation of polluted water for agricultural use. *Ph.D. Thesis*, Department of Environmental Studies, Institute of Graduate Studies and research, Alexandria University, Alexandria, Egypt 3:195208

4. Haande, S., T. Rohrlack, and R. P. Semyalo. 2010. Phytoplankton dynamics and cyanobacterial dominance in Murchison Bay of Lake Victoria (Uganda) in relation to environmental conditions. *Limnologica* 41:20–29

5. Boomiathan, M. 2005. Bioremediation studies on dairy effluent using cyanobacteria. *Ph.D. Thesis,* Bharathidasan University, Tiruchirapalli, Tamil Nadu, India

6. Semyalo, R. P. 2009. The effects of cyanobacteria on the growth, survival and behaviour of a tropical fish (Nile Tilapia) and zooplankton (Daphnia lumholtzi). *Ph.D. Thesis,* University of Bergen, Norway

7. Whitton, B. A. and M. Potts. 2002. *The Ecology of Cyanobacteria: Their Diversity in Time and Space.* New York: Kluwer Academic Publishers

8. Schopf, J. W. 2006. Fossil evidence of Archaean life. *Phil Trans Royal Soc Lond B Biol Sci* 361:869–885

9. Stal, L. J. 2003. Nitrogen cycling in marine cyanobacteria mats. *Fossil and Recent Biofilms: A Natural History of Life on Earth,* W. E. Paterson, D. M. Zavarzin, and G. A. Krumbein (eds), pp. 119–139. Dordrecht: Kluwer Academic Publishers

10. Rios A de los, M. Grube, L. G. Sancho, and C. Ascaso. 2007. Ultrastructural and genetic characteristics of endolithic cyanobacterial biofilms colonizing Antarctic granite rocks. *FEMS Microbiol Ecol* 59:386–395

11. Gorbushina, A. A. 2007. Life on the rocks. *Environ Microbiol* 9:1613–1631

12. Berner, R. A. 1990. Atmospheric carbon dioxide levels over phanerozoic time. *Science* 249:1382–1386

13. Berner, R. A. 2006. GEOCARBSULF: A combined model for Phanerozoic atmospheric O_2 and CO_2. *Geochim Cosmochim Acta* 70:5653–5664

14. Badger, M. R., D. Hanson, and G. D. Price. 2002. Evolution and diversity of CO_2 concentrating mechanisms in cyanobacteria. *Funct Plant Biol* 29:161–173

15. Badger, M. R. and G. D. Price. 2003. CO_2 concentrating mechanisms in cyanobacteria: Molecular components, their diversity and evolution. *J Exp Bot* 54:609–622

16. Raven, J. A. 2003. Carboxysomes and peptidoglycan walls of cyanelles: Possible physiological functions. *Eur J Phycol* 38:47–53

17. Solomon, S., D. Qin, M. Manning, Z. Chen, M. Marquis, K. Averyt, M. Tignor, and H. Miller. 2007. *Contribution of Working Group I to the Fourth Assessment Report of the Intergovernment Panel on Climate Change.* Cambridge, UK: Cambridge University Press

18. Schenk, P. M., S. R. Thomas-Hall, E. Stephens, U. C. Marx, J. Mußgnug, C. Posten, O. Kruse, and B. Hankamer. 2007. Second generation biofuels:

High-efficiency microalgae for biodiesel production (PUB-Publications at Bielefeld University). *Bioenergy Research* 1:20–43

19. Stainforth, D. A., T. Aina, C. Christensen, M. Collins, N. Faull, D. J. Frame, J. A. Kettleborough, S. Knight, A. Martin, J. M. Murphy, C. Piani, D. Sexton, L. A. Smith, R. A. Splcer, A. J. Thorpe, and M. R. Allen. 2005. Uncertainty in predictions of the climate response to rising levels of greenhouse gases. *Nature* 433:403–406

20. International Energy Outlook. 2008. What will it take to stabilize carbon dioxide concentrations? DOE/EIA-0484. Details available at http://www.eia.doe.gov/oiaf/ieo/ scdc.html (02.16.09)

21. Herzog, H., B. Eliasson, and O. Kaarstad. 2000. Capturing greenhouse gases. *Sci Am* 282:72–79

22. Herzog, H., K. Caldeira, and J. Reilly. 2003. An issue of permanence: Assessing the effectiveness of temporary carbon storage. *Clim Change* 59:293–310

23. Möllersten, K., J. Yan, and J. R. Moreira. 2003. Potential market niches for biomass energy with CO_2 capture and storage-opportunities for energy supply with negative CO_2 emissions. *Biomass Bioenergy* 25:273–285

24. Chisti, Y. 2007. Biodiesel from microalgae. *Biotechnology Advances* 25:294–306

25. Chisti, Y. 2008. Biodiesel from microalgae beats bioethanol. *Trends in Biotechnology* 26:126–131

26. Benemann, J. R. 1997. CO_2 mitigation with microalgae systems. *Energy Convers Manage* 38:S475–S479

27. Pulz, O. and W. Gross. 2004. Valuable products from biotechnology of microalgae. *Appl Microbiol Biotechnol* 65:635–648

28. Rodolfi, L., G. C. Zittelli, N. Bassi, G. Padovani, N. Biondi, G. Bonini, and M. R. Tredici. 2009. Microalgae for oil: Strain selection, induction of lipid synthesis and outdoor mass cultivation in a low-cost photobioreactor. *Biotechnol Bioeng* 102:100–112

29. Romero, J. M., F. G. Acién, and J. M. Fernández-Sevilla. 2012. Development of a process for the production of l-amino-acids concentrates from microalgae by enzymatic hydrolysis. *Bioresour Technol* 112:164–170

30. Oswald, W. J. and C. G. Golueke. 1960. Biological transformation of solar energy. *Adv Appl Microbiol* 2:223–262

31. Burnat, M., E. Diestra, I. Esteve, and A. Sole. 2009. In situ determination of the effects of lead and copper on cyanobacterial populations in microcosms. *Plos one* 4:6204

32. Kiran, B., A. Kaushik, and C. P. Kaushik. 2008. Metal-salt co-tolerance and metal removal by indigenous cyanobacterial strains. *Process Biochem* 43:598–604

33. Pereira, S., A. Zille, E. Micheletti, P. Moradas-Ferreira, R. De Philippis, and P. Tamagnini. 2009. Complexity of cyanobacterial exopolysaccharides: Composition, structures, inducing factors, and putative genes involved in their biosynthesis and assembly. *FEMS Microbiol Rev* 33:917–941

34. Ozturk, S., B. Aslim, and Z. Suludere. 2009. Evaluation of chromium (VI) removal behaviour by two isolates of *Synechocystis* sp. in terms of exopolysaccharide (EPS) production and monomer composition. *Bioresour Technol* 100:5588–5593

35. Sutherland, I. W. 2001. Biofilm exopolysaccharides: A strong and sticky framework. *Microbiology* 147:3–9

36. Vanhaverbeke, C., A. Heyraud, and K. Mazeau. 2003. Conformational analysis of the exopolysaccharide from Burkholderia caribensis strain MWAP71: Impact of the interaction with soils. *Biopolymers* 69:480–497

37. Otero, A. and M. Vincenzini. 2003. Extracellular polysaccharide synthesis by Nostoc strains as affected by N source and light intensity. *J Biotechnol* 102:143–152

38. Hu, C., Y. Liu, B. S. Paulsen, D. Petersen, and D. Klaveness. 2003. Extracellular carbohydrate polymers from five desert soil algae with different cohesion in the stabilization of fine sand grain. *Carbohydr Polym* 54:33–42

39. Arias, S., A. del Moral, M. R. Ferrer, R. Tallon, E. Quesada, and V. Bejar. 2003. Mauran, an exopolysaccharide produced by the halophilic bacterium Halomonas maura, with a novel composition and interesting properties for biotechnology. *Extremophiles* 7:319–326

40. Mancuso Nichols, C. A., J. Guezennec, and J. P. Bowman. 2005. Bacterial exopolysaccharides from extreme marine environments with special consideration of the southern ocean, sea ice and deep-sea hydrothermal vents: A review. *Mar Biotechnol* 7:253–271

41. Micheletti, E., S. Pereira, F. Mannelli, P. Moradas-Ferreira, P. Tamagnini, and R. De Philippis. 2008. Sheathless mutant of the cyanobacterium *Gloeothece* sp. strain PCC 6909 with increased capacity to remove copper ions from aqueous solutions. *Appl Environ Microbiol* 74:2797–2804

42. Abed, R. M. M., S. Dobretsov, and K. Sudesh. 2009. Applications of cyanobacteria in biotechnology. *J Appl Microbiol* 106:1–12

43. Parikh, A. and D. Madamwar. 2006. Partial characterization of extracellular polysaccharides from cyanobacteria. *Bioresour Technol* 97:1822–1827

44. Ghosh, T., K. Chattopadhyay, M. Marschall, P. Karmakar, P. Mandal, and B. Ray. 2009. Focus on antivirally active sulfated polysaccharides: From structure-activity analysis to clinical evaluation. *Glycobiology* 19:2–15

45. Bala, K., A. Kaushik, and C. P. Kaushik. 2007. Biosorption of Cr (VI) by native isolate of Lyngbya putealis (HH-15) in the presence of salts. *J Haz Mat* 141:662–667

46. Haszeldine, S. R. 2009. Carbon capture and storage: How green can black be? *Science* 325:1647–1651

CHAPTER 14
Production of Multi-purpose Fuels through Carbon Capture and Sequestration

V. K. Sethi

Director, UIT and Head, Energy Department,
Rajiv Gandhi Technological University, Bhopal, India
E-mail: vksethi@rgtu.net

INTRODUCTION

Energy security and energy independence are the prime concerns of our time as they will help in transforming India into a knowledge superpower of the world. The Indian power sector has taken rapid strides and reached an installed capacity of over 212,000 megawatt (MW) in 2012. It now has plans to add another 100,000 MW within the next five years. Global concern for reduction in emission of greenhouse gases (GHGs), especially carbon dioxide (CO_2) emissions, is likely to put pressure on the Indian power system to adopt not only improved generation technologies but also low carbon/no carbon power generation technologies [1]. In the power sector, we have already started witnessing a transition from conventional power generation technologies to green power technologies. A major thrust on CO_2 reduction on long-term and sustainable basis would come through adoption of advanced technologies of power generation such as Supercritical/Ultra-supercritical Power Cycles, Integrated Gasification Combined Cycles (IGCC), Fluidized Bed Combustion/Gasification Technologies, Renewable Energy Technologies, Biofuels, and other such green energy technologies [2]. In the short term, the focus is on energy conservation measures and use of selected hybrid of 'renewable' such as solar–wind–biomass.

In the long-term, while India is committed to developing all the future power plants based on supercritical technology and IGCC is also on the anvil of reducing the emissions and increasing the energy efficiency, the process of recycling CO_2 would be a value addition and a major technological breakthrough that will help directly reduce the impact of GHG emissions even in the supercritical power systems. CO_2 when recycled either to produce carbon monoxide (CO) or methanol (CH_3OH) or to be reused as a fuel would also release CO_2 into the atmosphere but would recycle CO_2 much the same way as the CO_2 from the biomass.

The gasification process would convert CO_2 into CO by a high temperature Boudard's Reaction using a carbon source. From 44 kg CO_2 and 12 kg carbon, 56 kg CO gets produced. The production of CO would increase as the recycle will result in large volumes of CO being produced; hence some or more than half of the CO needs to be removed from the system. This can be done by converting this CO into hydrogen (H_2) and CH_3OH, and in the process generating energy credits from the recycle of CO_2 which is a waste. This energy credit will lower the overall energy consumption in the power cycle and, thereby, increase the energy efficiency and lower the emissions in the power cycle.

When flue gas stream is recycled by converting the same into a useful fuel, it adds another cycle or a tertiary cycle viz., a gasification cycle of conversion of CO_2 to CO, the later being a useful fuel for recycling into the boiler. This process will again result in energy gain and lower GHG emissions from the power plant. This is also called flue gas regeneration cycle. It will result in the co-production of CH_3OH, a stable alcohol, that is also suited for delivering H_2 fuel for transportation. In this process, GHG emissions will get reduced in the transport sector and the fleet energy efficiency of the automobile fleet will get increased.

The chemistry of CO_2 recycle is as follows:

(i) Capture of CO_2 and its recycle can increase energy efficiency and, thus, reduce emissions.

(ii) The CO_2 capture follows an oxy-fuel combustion system where the emission of the power plant is 100% CO_2 while those of other coal-based power plants have 12%–15% CO_2 in the flue gas stream requiring costly capture processes.

(iii) The CO_2 is then converted into CO using a carbon source and the waste thermal heat of the boiler/gasifier. The gain in energy is 8.5% in terms of overall energy inputs.

(iv) This process of twin gasification is used to convert the entire coal fuel feed stock into clean coal gas and the H_2 in this feed gas is converted into CH_3OH. This will also obviate the need for coal quality control.

(v) The remaining coal gas, mainly CO, goes to the supercritical boiler for burning with oxygen as fuel.

(vi) As the feed gas is purified, the quantum of CO_2 to be used for recycle and reduction would depend on the emission regulations at that point of time.

The energy credits for different power cycles are discussed here.

(i) CO_2 recycling for conversion to CO gives 21.88% heat gain.

(ii) Producing H_2 from recycled CO_2, the gain in the input heat would be 18.69%.

(iii) Producing CH_3OH from recycled CO_2, there is a net gain in heat value in terms of 10.65%.

The conversion processes are explained in Appendix I.

STUDY CONDUCTED AT RAJIV GANDHI TECHNOLOGICAL UNIVERSITY

The quest for energy independence of the Rajiv Gandhi Technological University (RGPV) campus through clean energy technologies led to the setting up of a Green Energy Technology Center (GETC). The focus areas of GETC is diverse and includes clean coal technologies and Clean Development Mechanism (CDM), renewable energy technologies, biofuels and biodiesel, energy conservation and management, and CO_2 capture and sequestration technologies. The university is committed in developing green power technologies that will help generate 1 MW power within the campus and also provide formidable support to the national mission on green power and sequestration of CO_2. The GETC also has a 100% producer gas-based biomass gasifier and biodiesel reactor.

The CO_2 capture technology is on the anvil of this university. To this end, a pilot plant is being developed that will capture CO_2 and convert it into useful fuels such as H_2 for fuel cell application, methane (CH_4) for CH_4-based turbine for power generation, and biodiesel through algae route. This impact project has been awarded by Department of Science and Technology (DST), Government of India for CO_2 sequestration and production of multi-purpose fuels such as CH_4, H_2, and biodiesel through algae route through a single pilot plant. The pilot plant is considered a major milestone in commercialization of this frontier technology of carbon capture and sequestration. The systems being provided include CO_2 sequestration system, catalytic flash reduction of CO_2 using charcoal from gasifier, production of H_2 from CO and production of CH_4 using catalytic converter processes and capture of CO_2 using solar flux and

CO_2 as an input to the algae culture. Algae yield is being used for biodiesel production.

OBJECTIVES OF THE STUDY

The study objectives are as follows:

(i) Establish scientific basis for dry gas CO_2 separation, adsorption, storage, and conversion into fuel.
(ii) Develop zero emission technology projects and recycle CO_2 to add value to clean energy projects by adopting three pathways:
   (a) Sequester CO_2 and convert the same into fuel molecule.
   (b) Use CO_2 to grow microalgae to produce biodiesel and CH_4 gas.
   (c) Use CO_2 sequestration to grow energy plantation.
(iii)Develop mathematical and chemical models for CO_2 sequestration, hydro gasifier, and algae pond systems.
(iv)Establish a pilot plant for CO_2 sequestration and conversion into multi-purpose fuel.

METHODOLOGY AND SYSTEM CONFIGURATION

The systems being incorporated include CO_2 sequestration system, catalytic flash reduction of CO_2 using charcoal from gasifier, production of H_2 from CO, and production of CH_4 using catalytic converters. A portion of the captured CO_2 along with solar flux shall be fed to the algae culture in the specially designed algae pond. Algae yield can then be used for biodiesel production. The project team is presently working on the CO_2 capture technology from flue gas. The team is also working on a new approach for the efficient conversion of CO_2 to multi-purpose fuels.

A novel concept of CO_2 capture, storage, sequestration, and conversion into multi-purpose fuels is thus being achieved by conversion of CO_2 to CH_4, CH_4 through catalytic process, CO_2 to H_2, H_2 through carbon shift, CO_2 to algae and algae to biodiesel through CO_2, and solar flux.

The system configuration is shown in Figure 1. The broad specifications of the system are as under:

(i) Rated capacity of the capture of CO_2: 500 kg/day
(ii) Source of CO_2: Boiler of capacity 100 kg/hr steam and biomass gasifier of 10 kW

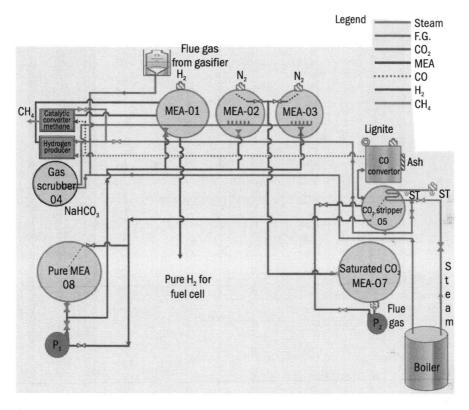

Figure 1 *Schematic diagram of CO₂ capture pilot plant*

(iii) Solvent used for capture of CO_2: Mono ethanol amine (MEA)

(iv) SO_x and NO_x removal: Scrubber unit $NaHCO_3$

(v) Catalytic converters/reduction unit
- For CH_4: input CO and H_2, specially developed catalyst "R-01"[*]
- For H_2: input CO and steam
- For CO: input CO_2 and lignite

RESULTS OF THE STUDY

In the first phase of testing of the plant covered, the following two test activities were carried out:

(i) Capture of CO_2 in MEA tank numbers 2 and 3 for the multiple CO_2 sources viz., boiler with diesel and biodiesel firing, and biomass

---

[*] Patent applied for

gasifier. A CO_2 capture level of 70%–78% was achieved for the above three post-combustion CO_2 sources.

(ii) Release of CO_2 from the stripper unit through steam coil heaters was carried out. The released CO_2 was diverted to the algae pond and the level of the released CO_2 was found to vary between 0.8% and 1.2%.

The results of the study are as follows:

(i) The pilot plant together with the combustion gas analyser and data acquisition system has been used for 4000 hours trail run for 'Uncertainty Analysis' in the experimentation. Once the trial run was completed, a CO_2 capture level of 90%–93% was achieved for the above post-combustion CO_2 sources, the boiler and a gasifier. It is seen that CO_2 capture of up to 93% has been achieved and H_2 formation is to the extent of 21%.

(ii) In-house development of catalysts for CH_4 is being attempted.

(iii) Lipid testing of CO_2 captured algae is being carried out on a variety of algae, and pond water analysis is being done and design modifications in the algae pond are being attempted.

(iv) Algal oil was extracted from algal biomass using soxhlet extraction by two different solvents viz., hexane and petroleum ether. Hexane extracted (4.76%) more oil from the algae than petroleum ether (2.52%) by soxhlet extraction procedures.

(v) The pilot plant installed at RGPV (Figure 2) is thus being utilized for a variety of applications during trial run of 4000 hours for process stabilization in the study of CO_2 capture in MEA ranging from 1 to 5 molar solutions; sequestration of CO_2 released from the stripper unit to a variety of algae and development of lipid content for biodiesel production; the pilot plant as well as table top plant is being used for development of low-cost catalysts for production of fuel elements such as H_2 and CH_4.

CONCLUSION

A pilot unit was constructed in the RGPV to produce multi-purpose fuel from CO_2 captured from fossil-fired power plants. The size of the plant is selected well above the lab scale unit so as to produce appreciable moles of CH_4 and H_2. It is estimated that a full-scale plant on a 500 MW pulverized coal-fired unit would require a plant of 510 tonnes/hour capacity.

Figure 2 *CO₂ sequestration plant (CO₂ and Steam)*
Source: Oil Fired Baby boiler

The simulation study has shown that if the technology of CO_2 capture recycling and sequestration is applied on a 500 MW coal-based thermal power plant with 30% capture, the following benefits are expected:

(i) Levelized cost of electricity (LCOE) on a long-term basis calculated through simulation exercise for retrofitting would be ₹1.05 per kWh. The energy penalty for 30% abatement would be 3% and the loss in generation due to use of steam in MEA process would be 15,000 kWh/hour for 30% CO_2 reduction. The capital cost would be ₹1.50 crore/MW.

(ii) The net emission reduction when the recycling of CO_2 is used in tandem with abatement would be 40% or down from 0.9 kgm CO_2/kWh to 0.54 kgm CO_2/kWh on a retrofitted thermal plant.

It is expected to resolve certain frontline issues in CO_2 sequestration such as energy intensive process optimization in terms of cost of generation and development of effective catalyst for CH_4, H_2, and biodiesel recovery through algae route.

Summary

Carbon dioxide capture and sequestration, and production of multi-purpose fuels, such as H_2, CH_4, and biodiesel through algae route in a post-combustion operation have been successfully demonstrated. A CO_2 sequestration pilot plant was constructed at the State Technological University of Madhya Pradesh (MP), 'RGPV'. This pilot project not only revalidated the possible use of amine absorption system to strip CO_2 from flue gases but also validated the data on its efficiency for a power plant. A CO_2 capture of over 93% has been achieved using MEA solvent of 20% concentration and the required heat for stripping the captured CO_2 is 3.88 MJ/kg of recovered CO_2, which is provided by the low pressure steam of about 150°C and 2 bar pressure from the associated boiler of 100 kg/h capacity. Although the stripper uses a low grade steam and some heat it contains was not used for generating power anyhow, it still causes 20% reduction in power output of the boiler. Using the water gas shift reaction and a lignite/charcoal gasifier, about 18% H_2 is being produced in this pilot plant, paving the way to the production of multi-purpose fuels from captured CO_2. Efforts are underway to produce CH_4 from stable CO and H_2 at elevated temperature, in a catalytic converter.

The CO_2 from the stripper unit is also diverted to an open algae pond, where solar flux is concentrated using parabolic collectors. A stimulation study of CO_2 capture and sequestration plant on an actual coal thermal power plant has been carried out.

ACKNOWLEDGEMENT

Author gratefully acknowledge the advice and support received from Professor Piyush Trivedi, Vice-Chancellor, RGPV; P. B. Sharma, Vice-Chancellor, DTU, Delhi; Professor Mukesh Pandey, Dean. The author is also thankful to his colleagues Ms Savita Vyas, Mr Pankaj Jain, and Mr Anurag Gour in helping prepare this paper.

REFERENCES

1. Reddy, D.N., K. Basu, and V. K. Sethi. 2003. IGCC for power generation: A promising technology for India, Chia Laguna, Sardinia, Italy. *CCT International Conference on Clean Coal Technologies for Our Future*, 21–23 October

2. Rao, J. S., N. V. S. Ramani, G. Srikanth, and J. Neelima. 2003. IGCC a clean coal technology for power generation in 21st century. *CCT International Conference on Clean Coal Technologies for Our Future,* 21–23 October

3. Reich, E. A. 2003. Chevron Texaco, coal gasification–the production of chemicals and clean electric power via the Integrated Gasification Combined Cycle (IGCC). *CCT International Conference on Clean Coal Technologies for Our Future,* 21–23 October

4. Juniper, L., D. Harris, and J. Patterson. 2003. Australian clean coal technology. *CCT International Conference on Clean Coal Technologies for Our Future,* 21–23 October

5. Meier, H. J. and K. Theis. 2003. Cleaner and technologies–a new initiative for Europe. *CCT International Conference on Clean Coal Technologies for Our Future,* 21–23 October

6. Proceedings of Indo-European Seminar on Clean Coal Technology and Power Plant Upgrading. 1997. Jan 14–16, CENPEEP, NTPC, Noida

7. TERI. 2005. *TERI Energy Data Directory and Year Book 2004–05.* New Delhi: TERI

8. Sharma, P. B. and V. K. Sethi. 2008. Mitigating climate change through green energy technologies. *International Seminar on New Horizons of Mechanical Engineering (ISME 2008),* March 18–20, at Rajiv Gandhi Technological University, Bhopal, M P

9. Sethi, V. K, Mukesh Pandey, and Savita Vyas. 2007. Production Potential of Electricity from Biomass in Indian context. *Ultra Science* 19(1): 1–10

10. Sethi, V. K., Mukesh Pandey, and Savita Vyas. 2010. A novel approach for CO_2 sequestration and conversion into useful multipurpose fuel. *International Journal ICER, JERAD*

APPENDIX

AN EXPLANATORY NOTE ON CARBON RECYCLING THROUGH CCS

MODE A: CO_2 RECYCLING FOR CONVERSTION TO CO

The process diagram is explained here

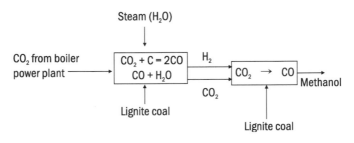

Energy in various molecules:

(i) CO_2 production is an exothermic reaction having energy $(-)393.5$ kJ/mol. There is no energy in this molecule after its formation and the value of the exhaust CO_2 is in fact zero. The CO_2 here in heat balance is seen as a waste, which it is.

(ii) H_2 has a heat value of 141.8 MJ/kg or the heat value would be 33,875 kcal/kg or in terms of power (1 kWh = 860 kcal) would be 39.40 kWh/kg.

(iii) Likewise CH_3OH has 22.7 MJ/kg. This would mean a heat value of 5423 kcal /kg or in terms of power it would be equivalent to 6.30 kWh/kg.

(iv) CO has a heat value of 10.112 MJ/kg. This would mean a heat value of 2416 kcal/kg or in terms of power it would be equivalent to 2.8 kWh/kg.

REACTION OF CO_2 TO CO

In the reaction of CO_2 to CO, we react 44 kg of CO_2 having a heat value of zero with 12 kg of carbon having a heat value of 7840 kcal/kg. The total heat value of input would be carbon 12 kg × 7840 = 94,080 kcal.

From this input we get 56 kg of CO having a heat value of 2416 kcal /kg or a total of 56 × 2416 = 135,296 kcal.

$135,296 - 94,080 = 41,216$

In percentage terms this amounts to an increase of 41,216/94,080 × 100 = 43.80%.

But the reaction is endothermic requiring heat input which must be accounted for through the ratio of heat output to heat input as under:

(i) Heat is required to convert CO_2 into CO and this has been calculated as the percentage of the input heat by means of the following reactions:

$C + CO_2 \rightarrow 2\ CO$

$\Delta H = 2 \times (-110.5) - (CO_2 = 393.5 - 221) = +172.5$ kJ, (Endothermic)   (1)

(ii) $2CO + O_2 \rightarrow 2CO_2$

$\Delta H = 2 \times (-393.5)$ kJ $= -787$ kJ, (Exothermic)   (2)

ΔH Balance for the two reactions: [787 − 172.5 = 614.5 kJ]

Or 614/787 = 78.08% heat output of the heat input.

Thus, endothermic heat as percentage of input heat value is

100 − 78.08 = 21.92%.

Thus, finally if we calculate the heat balance in terms of input heat value, we have to deduct the heat value equal to 21.92% from the input heat value of 94,080 kcal and this would be 20,622 kcal.

When we deduct this from the gain in heat value of 41,216 kcal, we still have a heat balance/gain of 41,216 − 20,622 = 20,594 kcal.

This would translate into

20,594/94,080 × 100 = 21.88% in terms of value of inputs.

This 21.88% is the heat gain if we produce only CO from CO_2.

MODE B: PRODUCING H_2 FROM RECYCLED CO_2

From basic chemistry, we know that

(i) Heat value of H_2 : 141.80 MJ/kg or 33,875 kcal/kg

(ii) Heat value of CH_3OH : 22.7 MJ/kg or 5425 kcal/kg

(iii) Heat value of CO : 10.112 MJ/kg or 2417 kcal/kg

(iv) Normally 1 kg of H_2 is produced commercially from 50 units of power electrolysis, and, hence, the heat value of input to produce 1 kg of H_2 would be 860 × 50 = 43,000 kcal/kg.

B (i)   Without recycling of CO_2, the following would hold good:

$CO + H_2O = CO_2 + H_2$ (3)

Molar weights would be 28 kg + 18 kg = 44 kg + 2 kg

In terms of heat values 28 × 2417 + 18 × 640 =

67,676 kcal + 11, 520 kcal = 79,196 kcal

For

2 kg H_2, the heat value would be 2 × 33,875 = 67,750 kcal
or 85.55% of input heat value or a heat loss of (–)14.55%.

B (ii)   When CO_2 is recycled there is a heat gain of 21.88%
or 21.88% of 67,676 = or a gain of 14,830 kcal in heat value
terms would be there.

Then heat value of recycled 28 kg of CO would be

67,676 × (100 – 21.88) = 52,846 kcal

The above reaction from recycled CO_2 in heat value terms would
be:

28 kg recycled CO = 52,846 kcal + 18 kg steam 11,520 kcal =
2 kg H_2.

Thus heat input is 52,846 kcal + 11,520 kcal = 64,366 kcal.

And 2 kg of H_2 would be produced and its normal heat value
would be 67,750 kcal denoting a heat gain in molar terms of
3384 kcal for 2 kg H_2.

The gain in the input heat would be 14,830/79,196* × 100
= 18.72%

In terms of normal inputs or a net heat energy gain of 18.72%.

MODE C: CH_3OH PRODUCTION FROM RECYCLED CO_2

When we produce CH_3OH from recycled CO_2 and H_2 is produced from
recycled CO, then this would result in the following heat gain:

C (i) CH_3OH synthesis in kcal when produced normally:

Formula $2H_2 + CO = CH_3OH$ (4)

or molar weight   4 kg + 28 kg = 32 kg

4 × 33,875 + 28 × 2417 = 203,176

Output $CH_3OH \rightarrow$ 32 × 5425 = 173,600 kcal

---

* This is because benefit or heat gain has to be related to normal heat input of 79,196 and not
with respect to input of recycled heat value of 64,366.

From an input heat value of 203,176 kcal input, we get 173,600 kcal as output.

The thermal efficiency of reaction will be denoted as 85.41% efficient.

C (ii) CH_3OH when produced from recycled CO_2, will lead to

(a) 4 kg of recycled H_2 would have a heat value of $2 \times 64,366 = 128,732$ kcal.

(b) 28 kg of recycled CO would have a heat value of $1 \times 52,846$ kcal.

Adding (a) and (b),

32 kg of CH_3OH from the above recycled H_2 and CO = 181,578 kcal.

Hence, 1 kg of CH_3OH produced from recycled H_2 and CO = 5674 kcal.

When CH_3OH is produced from normally heat input = 6351 kcal.

When using recycled inputs there is a net gain in heat value terms of 14.58%.

CHAPTER 15

CO_2 Storage and Enhanced Oil Recovery

Gautam Sen

Former Executive Director, Oil and Natural Gas Corporation
and
Senior VP Geosciences, Reliance Industries Ltd
B-341, C R Park, New Delhi-110019
E-mail: gautamsenindia@yahoo.com

INTRODUCTION

Carbon dioxide (CO_2) enhanced oil recovery (EOR) technologies have been commercially deployed in old oilfields mainly in the United States of America [1] for the last three decades. A pipeline network of CO_2 and identification of CO_2 sink that is, an old oilfield or saline aquifer in the sub-surface, if storage is for climate change, is necessary for deployment of this technology. Ability to produce incremental oil and therefore enhanced revenue through injection of high pressure CO_2 gas in producing field has also been tried and is financially attractive.

Sedimentary basins across the globe are the habitat for oil and gas fields, saline aquifers, and coal seams. The basin is a depocentre in the earth's crust formed due to tectonic forces either in extensional or contractional or in strike slip regime. Sedimentary rocks transported from highlands by water system deposit in this basin and the process continues with increased subsidence either by gravity or by renewed tectonic activity. In situ, organic growth of carbonate reefs is also possible under favourable environment. Sandstones, carbonates rocks usually have pores and they are either filled with saline aquifers or in some cases hydrocarbons. Coal also gets deposited in sedimentary basins and the coal seams can entrap fluids, particularly gases. It is important that this fluid does not escape back into the atmosphere and, therefore, top and lateral seals are necessary and usually impermeable rocks like shale/salts/evaporate or tight limestone provides the seal. Depleted oilfields and saline aquifers thus provide habitat for injected CO_2 in a sedimentary basin. Seismic data provide the sub-surface image of sedimentary basin and can indicate possible habitats for injecting CO_2.

A recent global study carried out using analogue from the USA experience indicate that over 50 large oil basins [2] have reservoirs amenable for miscible CO_2 injection, which can potentially produce 470 billion barrels of additional oil and store 140 billion metric tonnes of CO_2. Half of these basins are in Africa and Middle East. Around 300,000 barrels of oil per day is being produced in the USA mainly in Permian basin located around Texas from EOR through 60 Mt of CO_2 injection of which 20% is from anthropogenic sources. An extensive CO_2 pipeline network exists and the sinks, that is candidate for injection, is well-defined, including replacements. The process of water alternating gas injection manages the supply–demand of CO_2 at individual field level and shuts off CO_2 injection in case supply exceeds demand. European Union, Canada, Australia, and even some of the BRICS countries are also following suit.

Saline aquifers can store much larger volumes of CO_2 due to its abundance, but the ability to produce incremental oil with an injection of CO_2 in a matured oilfield offsets the costs associated with carbon capture and storage. Additionally, the geology of oil reservoirs is well-understood unlike reservoirs with saline aquifers, where leakage of CO_2 back into the atmosphere can pose a serious problem, unless the trapping mechanism exists. However, estimating performance of injected CO_2 in EOR is complex and data intensive, and each oilfield could be very different.

The EOR is a tertiary method after natural depletion and water injection methods have been applied to the oilfield. CO_2 is injected into an oil bearing stratum under high pressure. Miscible CO_2 interacts with the reservoir oil resulting in low viscosity, low interfacial tension, and, therefore, higher mobility. This, therefore, reduces the residual oil saturation. The challenge is to increase the area of contact of the two phases. In the case of immiscible CO_2 with reservoir oil, due to low reservoir pressure or heavy or too light oil, mobility still increases due to oil phase swelling and viscosity reduction of the mixture in some instances. CO_2 produced along with hydrocarbon, particularly in miscible case, is separated and recycled back into the oil reservoir.

Recent changes in EOR technology incorporate very large volume of injected CO_2 [1–1.5 hydrocarbon pore volume (HCPV)], better mobility control through alternate use of gas and water, advanced and optimized infill well drilling and completion to target reservoir above oil–water contact, and bypassed oil zones left unswept during the secondary water injection stage. Rigorous monitoring of reservoir fluid either through pressure studies in slim hole for thinner reservoirs or time-lapse seismic for seismically resolvable reservoirs is a significant approach in recent times. There is also a possibility of increasing geological storage of CO_2 in the saline aquifer once oil has been produced. Industrializing the process would include a steady supply of CO_2 through pipelines and selection of large number of oilfields as sink candidates including identification of their replacements, when oil production from a field becomes financially unviable.

CASE STUDIES FROM USA AND CANADA*

The Permian basin of south-west Texas and south-east New Mexico [3] is one of the largest and the most active oil basins in the USA, with the entire basin accounting for approximately 17% of the total USA oil production, with OXY as the main operator. Approximately two-third of OXY's Permian basin oil production is from fields that actively employ CO_2 flooding, an EOR technique in which CO_2 is injected into oil reservoirs, causing the trapped oil to flow more easily and efficiently. OXY is an industry leader in applying this technology, which can increase ultimate oil recovery by 15% to 25% in the fields, where it is employed in the Permian. Each year, OXY injects more than 550 billion cubic feet of CO_2 into oil reservoirs in the Permian, making OXY the largest injector of CO_2 for EOR in the USA and a world leader in this technology.

To ensure an adequate supply of CO_2 for EOR operations in the Permian, OXY relies upon several sources. These include CO_2 produced from the OXY-operated Bravo Dome field in north-eastern New Mexico and additional supplies from methane (CH_4) fields in the south-western Permian. The Century Plant in Pecos County, Texas further expands OXY's EOR infrastructure in the Permian basin. The plant processes natural gas with high-CO_2 content, resulting in CH_4 gas from the market as well as a major new source of CO_2 for OXY's Permian operations.

The principal reservoirs are carbonate shelf and platform strata with variable porosity and low permeability. Primary recovery efficiencies were low between 10% and 20% of OOIP (original oil in place). Water flooding started in the 1950s and CO_2 injection commenced in large fields in the 1970s. Recovery has gone up to 50% in some of the large

---

* Source: Global Technology Roadmap

fields with CO_2 injection. A recent USGS (United States Geological Survey) analysis indicates that the largest reservoirs – the San Andres, Grayburg, and Canyon – have contributed to this growth in oil reserves in the preceding decade. Deeper and smaller reservoirs can also be taken up, but with lower gain. The main reservoirs studied by USGS have a potential to further add a mean of 2.68 BBO of reserves through water flooding and CO_2 injection.

The Weyburn oilfield [4], operated by EnCana, Canada's largest oil company, is 130 km (80 mi) south-east of the city of Regina in Saskatchewan province. The Weyburn oilfield was discovered in 1954 with an estimated 1.4 billion barrels of original oil in place. Oil production started in 1955 and rose to about 31,500 barrels of oil per day in 1963. Starting in 1964, water was pumped into injection wells in order to increase oil production. By 1966, production peaked at about 47,200 barrels per day (Figure 1). Over the next 20 years, production declined steadily, dropping to just 9400 barrels per day by 1986. Additional vertical and horizontal wells were drilled. This increased production to approximately 22,000 barrels per day.

By 1998, roughly 330 million barrels of oil had been produced. This amounted to about 23% of the oil in the reservoir. Production again started declining rapidly. It was predicted that unless a new solution could be found to enhance oil recovery, total production would be no more than 350 million barrels, just 25% of the original oil in place.

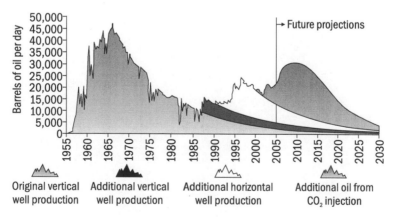

Figure 1 *EOR and future predictions from Weyburn oilfield*

The Great Plains Synfuels Plant

To encourage the development of alternative fuel sources, the USA Government supported the building of the Great Plains Synfuels Plant near Beulah, North Dakota. Commercial operations started in 1984. The goal was to produce CH_4 from coal. Everyday, more than 16,000 tonnes of crushed lignite coal are fed into "gasifiers", where it is mixed with steam and oxygen and then partially burned at a temperature of 1200°C (2200°F). This breaks down the coal to produce a mixture of gases. The gas is cooled to condense tar, water, and other impurities. Then, it is passed through methanol at −70°C (−94°F). This separates the synthetic natural gas (SNG), mostly CH_4, from other compounds mostly CO_2.

The daily production is 3050 tonnes of SNG, which is fed through gas pipelines to customers and 13,000 tonnes of waste gas, 96% of which is CO_2. Many synfuels plants release their waste gas into the atmosphere, contributing to the greenhouse gas effect and global warming. Waste gas from the Great Plains Plant is fed into a 330 km (205 mi) pipeline to Weyburn, where it not only is disposed of safely, but also helps to produce more oil (Figure 2).

The Weyburn CO_2 EOR Operation

In 1997, the Dakota Gasification Company (DGC) agreed to send all of the waste gas (96% CO_2) from its Great Plains Synfuels Plant through a pipeline to the Weyburn oilfield.

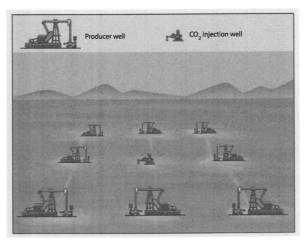

Figure 2 *Well grid structure in Wyeburn project*

Delivery of the first CO_2 to Weyburn commenced in September 2000. The gas in the pipeline is at a very high pressure (about 152 bar), which makes it a supercritical fluid. Supercritical fluids are gases under such high pressures that the vapour (gas) phase becomes as dense as the liquid phase. Supercritical fluids have high density, but they flow easily like gases and so are ideal for transporting through pipelines. The Weyburn oilfield has a total of 720 wells. The vertical wells were drilled in a "9-spot" grid pattern–eight producing wells in a square around an injection well and typically having a spacing of around 150 m (500 ft). The high-pressure CO_2 is being pumped into 37 injection wells, helping oil to flow towards 145 active producer wells (Figure 3).

The level of purity of the CO_2 supplied is ideal for use in EOR. This is because CO_2 dissolves more readily into oil, when small impurities are present. Hydrogen sulphide (H_2S), which makes up 2.5% of the injection gas, is particularly beneficial in helping CO_2 to mix with oil.

When CO_2 supercritical fluid is pumped at high pressure into the reservoir, the CO_2 mixes with the oil, causing it to swell and become less viscous (Figure 3). The swelling forces oil out of the pores in the rocks, so that it can flow more easily. Water is pumped into the injection wells, alternating with CO_2, to push the released oil towards producer wells. Some CO_2 comes back out of the ground at producer wells. This is recycled, compressed, and reinjected along with gas from the pipeline.

It is predicted that the CO_2-EOR operation will enable an additional 130 million barrels of oil to be produced, extending the field's commercial life by approximately 25 years. It is also anticipated that about 20 Mt

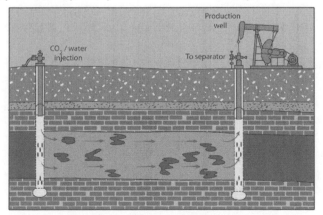

Figure 3   *CO_2 injection and oil recovery process*

of CO_2 will be injected and become permanently stored 1400 m (4600 ft) underground over the lifetime of this project. There is worldwide interest in this test of the viability of underground storage for large-scale reduction in CO_2 emissions to the atmosphere. The Weyburn CO_2 Monitoring and Storage Project is funded by several international energy companies, the USA and Canadian Governments, and the European Union. The main concern is whether the CO_2 will stay in place.

Weyburn is an excellent test site because, since 1955, thorough geological tests have been made and the results stored. There are rock core samples from 1200 boreholes, plus time lapse seismic analysis and borehole logging. Researchers are also sampling groundwater to test for CO_2 leaks in wells. So far, no leaks have been detected and none of the gas has escaped to the surface. The Canadian Government believes that CO_2 storage deep underground will help it to meet its targets under the 1997 Kyoto Protocol, which requires a reduction in greenhouse gas emissions by an average of 5% between 2008 and 2012. In the case of Weyburn, the CO_2 is from coal that came from under the ground, so it is effectively being put back, where it came from.

Another CO_2 capture and storage project currently underway is in the Sleipner field [5] in the North Sea in the Norwegian region. Injection commenced in 1996 with CO_2 separated from natural gas and injected into the Utisara sand, a saline aquifer of late Cenozoic age at a depth of 1012 m below sea bed, around 200 m below the reservoir top with a storage of over 11 Mt. Time-lapse surveys have been carried out with three-dimensional surveys in 1994, 1999, 2001, 2002, 2004, 2006, 2008, and so on. The CO_2 plume is seen in seismic as a number of sub-horizontal reflectors of tunes wavelets arising from thin layers of CO_2 trapped beneath thin intra-reservoir mudstone and the reservoir cap rock and growing in time. There is also a velocity pull down in reflectors below the plume. A post-stack stratigraphic inversion of the 1994 and 1996 provided P-wave impedance. Pre-stack inversion of the 1994 and 2006 data sets (after injecting 8.4 Mt) with 50 iterations in the window 750–1400 ms provided shear wave impedance and refined P or body wave impedance. These results when combined with spectral decomposition gave an idea of the thickness of CO_2 layers, thus making it possible to make a volumetric estimate of CO_2 in the sub-surface.

MONITORING/FLOW SURVEILLANCE

Acoustic and elastic impedance contrast between the overlying and underlying rock strata and the reservoir rock provides the basis for sub-surface imaging of the reservoir. While the longitudinal wave (P wave) is sensitive to the pore fluid, the transverse wave (shear wave) bypasses the pore and therefore is insensitive to the presence of fluid. Gassman's equation provides a relationship between P wave velocity (Vp), S wave velocity (Vs), the average density of matrix and fluid, porosity and elastic constants of rock matrix and fluid. Fluid substitution assuming constant shear modulus and constant porosity can help in modelling the change in P wave velocity with injected CO_2.

Laboratory studies along with well-calibrated seismic parameters and using Gassman's equation can provide sensitivity analysis of seismic velocity (P wave) with pressure and saturation of injected CO_2. Thus, with a baseline three-dimensional survey prior to CO_2 injection and then with three-dimensional survey repeated periodically along with injected CO_2 and production and interpreting changes in sub-surface as per the template designed as above, calibrated with saturation logs in time-lapse mode, it is possible to quantify enhance oil recovered, both above the oil–water contact and in bypassed zones in isolated pools. Additionally, inversion of pre-stack seismic data composed of reflections from different offsets can throw further light into the change in the sub-surface with injected CO_2 and enhanced production.

Reflection coefficient of non-zero offset seismic wave at an interface depends on the contrast in acoustic impedance along with the contrast in Poisson's ratio (which is essentially a function of the ratio of P wave and shear wave velocity) of the overlying and underlying bed. Thus, AVO (amplitude versus offset) behaviour of the reservoir also changes with injected CO_2 and production of oil. In case this behaviour is calibrated with well data, the changes could help in understanding fluid movement based on rock physics modelling.

However, this is easier said than done for artefacts in acquisition of seismic data and processing of seismic data can easily mar the assumption of repeatability. Three factors namely earth effects, acquisition-related effects, and noise change seismic amplitudes.

Earth effects [6] include spherical divergence, absorption, transmission losses, interbed multiples, converted waves, tuning, anisotropy, and structure. Acquisition-related effects include source and receiver arrays and receiver sensitivity. Noise can be ambient or source generated, coherent or random. Processing attempts to alleviate problems of non-repeatability, but can, in the process, create amplitude distortion and even now algorithms used are still not without their limitation in preserving amplitudes. Three-dimensional surveys done at different time are processed with the same processing parameters and the same sequence. At each step in the non-reservoir zone, data is matched through time and space variant amplitude and spectral trace matching after correcting for time shifts, if any due to variation in near surface between the surveys.

This process can reasonably ensure that at the non-reservoir level, the time-lapse surveys match and differences over and above this at the reservoir level can only be ascribed as signal attributed to injected CO$_2$ and enhanced production. Quantification of enhanced production and the economics can then only be on a firmer footing. Disadvantages of carbon sequestration through EOR vis-à-vis injecting CO$_2$ in saline aquifers are geographic distribution of old oil fields and their limited capacity. In addition subsequent burning of the additional oil so recovered will offset much or all of the reduction in CO$_2$ emissions. However vertical shift of flat spots, if seen in repeated 3-D seismic data, due to vertical movement of oil–water contact, can help in monitoring carbon sequestration, along with enhanced recovery in old oilfields.

The discussion so far assumed that CO$_2$ is inert [7] and it does not affect the rock matrix. However, this is not always the case and its effect also needs to be discussed. Temperature and pressure conditions in sub-surface will keep CO$_2$ in a liquid state. Under this condition if CO$_2$ replaces water without dissolving in it, the bulk modulus and density would decrease. When supercritical CO$_2$ migrates upwards, it moves to subcritical state and this may alter the free and dissolved gas ratio, affecting fluid saturation and distribution of patchy saturation in a mostly homogeneous saturation. Changes in P wave velocity with an increase in saturation of CO$_2$ are very different for the two cases below 70° saturation. Reduction in velocity of the P wave is much faster for homogeneous saturation as compared to patchy saturation and therefore unless lab studies of the core provide the type of saturation data, interpretation of lowering of Vp to saturation of CO$_2$ directly can be vastly erroneous. A 3% enhancement in porosity due to dissolution of mineral in CO$_2$ causes the same reduction in Vp as 70% enhancement in

CO_2 saturation. Thus, non-inert CO_2 is a difficult case for a quantitative estimation of injected CO_2.

Shear wave data may help to differentiate dissolution from saturation effects. With the injection of CO_2, there is an increase in Vs due to lowering of density in the inert case, while chemical dissolution causes a decrease in Vs due to lowering of rigidity modulus. Also the behaviour of Vs is independent of the type of saturation, patchy vis-a-vie uniform for shear wave bypasses fluid. The challenge is to acquire good quality shear wave data, particularly in four-dimensional sense. Combining both Vp and Vs particularly using time-lapse survey, backed up with extensive core data can provide a better quantitative estimation. Vertical seismic profiling that is, VSP carried out in three-dimensional sense can provide a higher resolution data since it involves only one-way wave propagation unlike surface seismic, which has two-way wave propagation. Static errors due to near surface irregularities are also much more pronounced in surface seismic data for shear wave acquisition.

We can conclude by stating that monitoring of multi-phase sub-surface flow associated with CO_2 injection poses new challenges, because of complex seismic response of CO_2 water rock systems. Conventional models for seismic signatures like Gassman's model based on fluid substitution are mechanical and cannot incorporate chemical changes to host rock, thereby altering porosity or the formation of patchy porosity on account of change of state from supercritical to subcritical. Thus, magnitude and sense of changes in Vp and Vs together, in time-lapse seismic/VSP backed up by laboratory studies of the core may help in quantitative estimates of injected CO_2 and each case can be unique depending on the interaction of CO_2 with water and host rock.

International Energy Agency has prepared a detailed compilation of the estimates of original oil in place, ultimate primary and secondary recovery, incrementally technically recoverable oil from CO_2-EOR and the volume of CO_2 stored in association with this process in 50 world basins with favourable conditions for miscible CO_2. Basins are located in the USA, Canada, Russia, China, Middle East, Africa, Latin America, and Australia. This process is cheaper in on-shore, but a large oilfield in off-shore like Brazil can be an ideal candidate in storing CO_2 in pre-salt reservoir.

CO_2 amounting to 139 Gt has been injected into the sub-surface with an average of 0.30 tonnes/bbl as the CO_2/oil ratio at the time of report in 2009. India needs to take up EOR methods more vigorously, though

ONGC has initiated the process. Gujarat, Assam, and Rajasthan can be the pioneers in this effort depending on the type of oil.

CONCLUSION

Capturing and storing CO_2 from atmosphere is essential for saving the planet in view of the large increase of CO_2 through anthropogenic source. This naturally upsets the normal carbon cycle. Sub-surface geology provides the best medium for storage of excess CO_2 either in old oil and gas fields or in saline aquifer. Depleted fields have an advantage for the geology is well known, the infrastructure exists, and the injected CO_2 can enhance recovery, thereby making the project financially attractive. It has to be ensured that injected CO_2 does not leak back to the atmosphere and entrapment mechanism is known for oil and gas fields unlike saline aquifer. However, the potential to store CO_2 is far greater in the saline aquifer and is ubiquitous.

Monitoring movement of injected CO_2 can be done by using time-lapse surface or down hole survey carried out in three-dimensional sense and both P wave and shear wave studies would be required to make a volumetric estimate. In case CO_2 is inert, the interpretation would be relatively simpler, else alteration of petro-physical characteristics of the host rock and the type of porosity data from extensive core studies using Scanning Electron Microscope would be required as input for interpreting time-lapse data. It can be concluded [8] that quantitative interpretation of repeated three-dimensional data for monitoring and volumetric computation of the stored CO_2 depends on the rock physics model. Vp, Vs, and amplitude changes with respect to CO_2 saturation in reservoir conditions of temperature and pressure need to be studied and each case is unique in its own way.

Summary

The EOR is a tertiary method after natural depletion and water injection methods have been applied to the oilfield. CO_2 is injected into an oil bearing stratum under high pressure. Miscible CO_2 interacts with the reservoir oil resulting in low viscosity, low interfacial tension, and, therefore, higher mobility. This, therefore, reduces the residual oil saturation.

Recent changes in EOR technology incorporate very large volume of injected CO_2 (1 to 1.5 HCPV), better mobility control through alternate *use of gas and water, advanced and optimized infill well drilling and completion to target reservoir above oil–water contact, and bypassed oil zones left unswept during the secondary water injection stage. Rigorous monitoring of reservoir fluid either through pressure studies in slim hole for thinner reservoirs or time-lapse seismic for seismically resolvable reservoirs is a significant approach in recent times.*

REFERENCES

1. National Energy Technical Laboratory. 2010. Carbon dioxide enhanced oil recovery. *US Department of Energy Report*
2. Godec, Michael L. *et al.* 2011. Global technology roadmap for CCS in industry sectorial assessment. *CO_2 Enhanced Oil Recovery Advanced Resources International Inc Report*
3. Tennyson, Marilyn E. *et al.* 2012. Assessment of remaining recoverable oil in selected major oilfields of Permian basin, Texas and New Mexico. *USGS Report*
4. Jinfeng, M. A. 2010. AVO modelling of pressure saturation effects in Weyburn CO_2 sequestration. *Leading Edge*. Littleton, CO: SEG Publication
5. Chadwick, Andy, *et al.* 2010. Quantitative analysis of time lapse seismic monitoring data at the Sleipner CO_2 storage operation. *Leading Edge*. Littleton, CO: SEG Publication
6. Bacon, M., R. Simm, and T. Redshaw. 2003. *3-D Seismic Interpretation*. Cambridge: Cambridge University Press
7. Tizana, Vanoria, *et al.* 2010. The rock physics basis for 4-D monitoring of CO_2 fate: Are we there yet. *Leading Edge*. Littleton, CO: SEG Publication
8. Avseth, Per, Tapan Mukherji, and Gary Mavko. 2008. *Quantitative Seismic Interpretation*. Cambridge: Cambridge University Press

CHAPTER 16

Real-Term Implications of Carbon Sequestration in Coal Seams

V. Vishal[1], T. N. Singh[2], S. P. Pradhan[2], and P. G. Ranjith[3]

[1]Department of Earth Sciences, Indian Institute of Technology Roorkee,
Uttarakhand-247667
[2]Department of Earth Sciences, Indian Institute of Technology Bombay,
Mumbai-400076
[3]Department of Civil Engineering, Monash University, Clayton, VIC-3800,
Australia
E-mail: vikram12july@gmail.com

INTRODUCTION

India is facing shortage in supplying electricity and clean cooking facilities to a majority of its population and, hence, infrastructural development is bound to happen in the years to come. Large consumption of available energy resources available markets and economic projections indicate that oil, gas, and coal will remain the mainstay for energy supply in the coming few decades. The burning of these fossil resources adds significant amounts of CO_2 in the atmosphere. Under a strict climate regime, the emissions of greenhouse gases (GHGs) into the atmosphere need to be controlled and prevent the rise in their concentration to limit global warming. One of the methods for reducing the atmospheric release of these gases is carbon sequestration. Geological sequestration is defined as the capture of CO_2 directly from anthropogenic sources and disposing it deep into the ground for geologically significant periods of time [1]. Coal seam sequestration as described by White et al. [2] is, "the storage of CO_2 from anthropogenic sources into deep, unmineable coal seams for geologically significant limits with or without concomitant recovery of natural gas."

Coal is both a source and a reservoir of natural gas; the volume of methane (CH_4) stored in each coal is dependent on many factors. CH_4 is retained in coal as adsorbed molecules on the organic micropores, free gas in pores, cleats and fractures, dissolved in solution within the coalbed. Injection of CO_2 in coal seams may be carried out to extract CH_4 through the fractures and cleats in the coal. The CO_2 has an additional effect compared to other gases that it is preferentially adsorbed onto coal surfaces, displacing CH_4 from adsorption sites. The CO_2 enhanced coalbed methane (ECBM) sequestration is a value addition project in the management of increasing atmospheric concentration of GHGs as it recovers the cost of capture, processing, transportation, and storage of CO_2 by production of CH_4.

Coal Seams: Unconventional Reservoir Behaviour

Challenges in coal seam sequestration are many folds. The developments in extracting natural gas from coal are relatively new, while the conventional reservoirs have been better understood due to several decades of their production history. Coal is structurally heterogeneous, and is composed of matrix blocks and natural fracture system called cleats. The flow characteristics of coal are variable in different directions that is, along the face cleats and butt cleats. The mechanism of storage of gas in coal is by physico-chemical adsorption, while in conventional reservoirs, the gases are stored in the pore spaces by mechanism of compression. The process of dewatering takes place in coal first and this leads to significant reduction of reservoir pressure. The reduced pressure in borewell favours desorption of gases from the coal seam. Hence, sequestration of CO_2 into deep unmineable coals and coalbed methane (CBM) producing coal seams needs to be carefully examined before implementation of the process.

Research work on field and laboratory scale is being carried out to understand the reservoir behaviour. Several pilot scale studies for injection of CO_2 is under operation in different fields such as Ishikari basin, Japan and North Dakota, USA. Most of these basins have experienced loss in injectivity due to reduction in coal permeability with time. The decline may be attributed to the CO_2 adsorption-induced swelling in coal mass, which leads to closure of macro-porosity (that is, cleats). The CO_2 in supercritical form induces maximum volumetric deformation in coal. Field test results from Williston basin coal seam in North Dakota showed that coal matrix swelling led to approximately 10 times decline in CO_2 permeability. A similar scenario was observed in Ishikari basin in Japan, where reduction of nearly 70% took place in the volumes of CO_2 from first year of injection to second year of injection. Nearly 35.7 metric tonnes of CO_2 could be injected during the first year at a rate of

2.3 tonnes/day without any gas production. However, the total amount dropped to around 11 metric tonnes in the second year of injection in Ishikari basin pilot project. However, the loss in injectivity was made up for by the decrease in effective stresses adjacent to injecting well. Similar observations were made in a few other coal basins such as Upper Silesian basin and San Juan basin, where CO_2 was being injected for ECBM recovery. Hence, it is important to understand the response of coal in detail before implementation of the sequestration process on a field scale.

CO_2 Sorption-induced Coal Matrix Deformation

Most of the gases are adsorbed into the micropores of coal, while some occur along the surface of cleats. Figure 1 shows a schematic of the matrix and cleats of coal. The adsorption/desorption of gases from coal matrix is associated with the deformation of the matrix. Adsorption of a given gas on a particular solid may be expressed as a function of pressure at a constant temperature called as adsorption isotherms. Adsorption isotherms are used to study the gas storage capacity of coal with respect to gas pressures (or concentrations) at a particular temperature. As the gases desorb from coal, the matrix undergoes shrinkage, while adsorption of gases by coal causes swelling of matrix. The "swelling" of coal due to gas adsorption is an increase in the volume occupied by coal as a result of the viscoelastic relaxation of its strained, glassy/rubbery and highly cross-linked macromolecular structure [3–6].

Coal has a higher affinity for CO_2 over CH_4 and, hence, larger amount of CO_2 is adsorbed than CH_4 which is desorbed from the coal matrix. Also, CO_2 induces higher coal matrix swelling as compared to CH_4. Several researchers have validated the adsorption of different phases by coal and quantified the resulting volumetric strain in coal samples of varying sizes from different basins [7–9]. Vishal *et al.* [10] quantified the amount of strain in Indian bituminous coal sample, when exposed to varying injection pressure of CO_2 in confined state. It is also observed that adsorption of CO_2 causes weakening of coal [11].

Swelling-related Loss in Permeability

Permeability in coal is estimated using the Darcy's law [12] for interpretation of experimental results, provided the volumetric flow rate varies linearly with pressure gradient across the ends of the sample. For higher flow rates, the pressure gradient may exceed from that predicted by

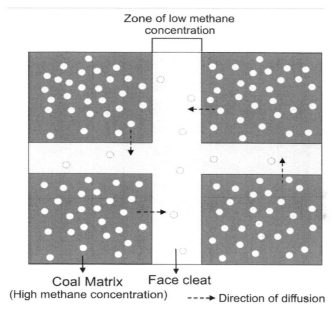

Figure 1 *Arrangement of cleats and matrix elements in coal*

Darcy's law and such behaviour is called as non-Darcy flow. Permeability of coal is maximum in the direction of face cleats. Gash [13] used coal samples from San Juan basin and obtained that permeability parallel to butt cleats (0.3 ~ 1.0 mD) and along the face cleats (0.6 ~ 1.7 mD) were comparable, while that vertical to bedding planes were two orders lesser (0.007 mD). As it was mentioned earlier, coal matrix swells due to adsorption of coal. This matrix swelling in turn closes the aperture of the fractures or the cleats and, hence, the path or conduits for passage of fluids and significant reduction in coal permeability is observed as a consequence. Coal exhibits differential sorption affinity to different gases and, hence, depending on the sorptive and non-sorptive gas type, the permeability of coal varies accordingly. Skawinski [14] showed that CO_2 resulted in still higher reduction in coal permeability as compared to CH_4 and nitrogen (N_2) due to high adsorption affinity of coal mass towards CO_2. Robertson [15] also demonstrated a decreasing order of coal permeability with N_2, CH_4, and CO_2, respectively. The major application of this phenomenon would be in enhanced recovery of CBM using CO_2 sequestration. Lin *et al.* [16] used Powder River basin coal and utilized pure N_2, CH_4, CO_2 and mixtures of N_2 and CO_2 under a constant effective stress. They established that coal permeability decreased with increase in gas sorption pressure, while increase in CO_2 component in the flowing mixtures reduced permeability to a greater extent. For pure gases, CO_2 caused highest permeability reduction followed by CH_4 and finally N_2.

Coal Seam CO_2 Sequestration in India

India is a coal-rich nation and has significant volumes of CBM reserves. The recent developments in CBM production show the positive way to further research on understanding the reservoir behaviour for carbon sequestration. India has the third largest proven coal reserves and the fourth largest producer of coal in the world. Moreover, the estimated CBM resources hold significant prospects for commercial recovery of natural gas. The total CBM resource in India has been estimated to be around 4.6 TCM. The deeper reserves where recovery of coal with or without CBM extraction seems less feasible, technologically as well as economically, may be the prime targets for sequestration of anthropogenic CO_2 for significantly large span of time. Coal also exists at depths, where economic mining may be limited. These blocks may be treated as a potential targets for CO_2 storage. A preliminary study by Halloway *et al.* [17] and IEAGHG [18] suggests that huge CO_2 storage potential exists in the major coalfields in India and for coals at depths above 1200 m, it could be of the order of 345 Mt that is approximately 6000 bcf of CO_2, while Mendhe *et al.* [19] estimated a total CO_2 storage capacity of 4459 Mt (equal to 80.26 tcf) in the Indian coalfields including both Gondwana and Tertiary coals. Power sector in India is mostly coal based and produces huge quantities of CO_2. There are several industries acting as point sources for CO_2. The close proximity of deeper coals along with the locations of point sources of CO_2 emission indicates high chances of sequestration in coal. Some of the basins have been marked with high sequestration potential of 1885 million tonnes (Mt) in the Cambay basin, while some like Singrauli offer only 1–2 Mt of storage depending upon the coal availability for sequestration and other parameters.

However, not many studies have been carried out on investigating the behaviour of Indian coal with CO_2 in terms of the real-term coal–CO_2 interactions. The role of effective stresses, confining stresses, coal matrix deformation and so on, on CO_2 permeability of Indian bituminous coal was found in a few recent research works reported by Vishal *et al.* [10, 21]. The results reveal that coal undergoes swelling and at higher depths, permeability reduces significantly, thereby altering the complete process of sequestration. Phase transformation of injected CO_2 at such depths may also lead to changes in coal–CO_2 interactions, finally influencing the injectivity of gases. Further, the influence of gas injection on the strength characteristics of coal as well as surrounding rock must be investigated for overall stability and safety of the system. Cap rock integrity is one of the most important parameters for sequestration of GHGs and must be understood in greater details before actual implementation.

Application of carbon capture and storage (CCS) in coal seams has its own advantages and, thus, focus on coal research for sequestration has gained attention over time. The mechanism of storage in coal being adsorption is safer as compared to the compressional mechanism in conventional hydrocarbon reservoirs. The preferred affinity of coal for CO_2 over CH_4 may be utilized for enhanced recovery of CBM from the coal blocks. Simulation works have indicated that injection of CO_2 may be used to recover more than 90% of CBM, which during primary recovery only enables upto 40%–50% of gas extraction. Vishal *et al.* [20] worked out the CO_2-ECBM recovery for a block of coal seam in Raniganj coalfield for a preliminary investigation of the process using numerical simulation. The results indicate a positive connotation for the feasibility of the process, while keeping in mind the assumptions and constraints in their work.

CONCLUSION

To meet the growing energy needs and maintain a cleaner atmosphere, CO_2 storage in coal is a promising opportunity, which provides value addition in terms of low carbon fuel that is enhanced CH_4 recovery. The process of enhanced CBM recovery shall also partly offset the cost of carbon capture, transportation and storage in coal and hence, falling in the category of carbon capture, storage and utilization. Due to the unconventional reservoir behaviour and specific challenges in coal seam sequestration, detail studies to establish a pilot project should be carried out in Indian basins. Understanding the behaviour of coal in underground scenarios is the key to successful operations and if established, coal seam sequestration could lead its way to GHG release mitigation in future.

Summary

Carbon sequestration in coal may be coupled with the enhanced production of CBM due to higher adsorption preference of coal for CO_2 over CH_4. The adsorption of CO_2 causes coal matrix to swell, which *leads to the closure of cleats or natural fractures. This closure of cleats results in reduced permeability in coal and, hence, the entire operation of coal seam sequestration may get adversely affected. Several field operations with pilot/demonstration projects have encountered reduced CO₂ permeability in coal due to swelling of coal matrix with time. Injection of CO₂ also leads to sorptive weakening in coal. It is important, therefore, to understand the response of coal to changing effective stresses, adsorption-induced coal matrix swelling, and the evolution of permeability due to these factors. For successful implementation of carbon sequestration in coal, detail reservoir characteristics should be understood and site-specific studies must be carried out.*

REFERENCES

1. Bachu, S. 2002. Sequestration of carbon dioxide in geological media in response to climate change: Road map for site selection using the transform of the geological space into the CO_2 phase space. *Energy Conver Manag* 43(1):87–102

2. White, C. M., D. H. Smith, K. L. Jones, A. L. Goodman, S. A. Jikich, and R. B. LaCount. 2005. Sequestration of carbon dioxide in coal with enhanced coalbed methane recovery: A review. *Energy Fuel* 19(3):659–724

3. Brenner, D. 1985. The macromolecular nature of bituminous coal. *Fuel* 64:167–173

4. Larsen, J. W., R. A. Flowers II, P. Hall, and G. Carlson. 1997. Structural rearrangement of strained coals. *Energy Fuel* 11:998–1002

5. Larsen, J. W. 2004. The effects of dissolved CO_2 on coal structure and properties. *Int J Coal Geol* 57:63–70.

6. Larsen, J. W. and J. Kovac. 1978. Polymer structure of bituminous coals. *ACS Symp Series* 71:36–49

7. George, J. D. and M. A. Barakat. 2001. The change in effective stress associated with shrinkage from gas desorption in coal. *Int J Coal Geol* 45:105–113

8. Mazumder, S., A. A. Karnik, and K. H. A. A. Wolf. 2006. Swelling of coal in response to CO_2 sequestration for ECBM and its effect on fracture permeability. *SPE J* 11(3):390–398

9. Perera, M. S. A., P. G. Ranjith, S. K. Choi, and D. Airey. 2011. The effects of subcritical and supercritical carbon dioxide adsorption-induced coal matrix swelling on the permeability of naturally fractured black coal. *Energy* 36(11):6442–6450

10. Vishal, V., P. G. Ranjith, and T. N. Singh. 2013. CO_2 permeability of Indian bituminous coals: Implications for carbon sequestration. *International Journal of Coal Geology*; doi: 10.1016/j.bbr.2011.03.031

11. Viete, D. R. and P. G. Ranjith. 2006. The effect of CO_2 on the geomechanical and permeability behaviour of brown coal: Implications for coal seam CO_2 sequestration. *Int J Coal Geol* 66:204–216

12. Darcy, H. 1856. *Les Fontaines publiques de la ville de Dijon.* Dalmont, Paris: Libraire des corps impériaux des ponts et

13. Gash, B. W. 1993. The effects of cleat orientation and confining pressure on cleat porosity, permeability and relative permeability in coal. *Abs. Log Anal* 33:176–177

14. Skawinski, R. 1999. Considerations referring to coal swelling accompanying the sorption of gases and water. *Arch Min Sci* 44(3):425–34

15. Robertson, E. P. 2005. Measurement and modeling of sorption-induced strain and permeability changes in coal. *Idaho Nat. Lab.* INL/EXT-06-11832

16. Lin, W., G. Q. Tang, and A. R. Kovscek. 2008. Sorption-induced permeability change of coal during gas-injection processes. *Reser Eval Eng* 11(4):792–802

17. Holloway, S., J. M. Pearce, V. L. Hards, T. Ohsumi, and J. Gale. 2007. Natural emissions of CO_2 from the geosphere and their bearing on the geological storage of carbon dioxide. *Energy* 32(7):1194–1201

18. IEAGHG. 2008. A regional assessment of the potential for CO_2 storage in the Indian subcontinent, Programme Report, *IEAGHG R & D Prog* Cheltenham.

19. Mendhe, V. A., H. Singh, Prashant, and A. Sinha. 2011. Sorption capacities for enhanced coalbed methane recovery through CO_2 sequestration in unmineable coal seams of India. *Int. Conf Unconv Sour Fossil Fuels Carb Manag.* Gujarat

20. Vishal, V., L. Singh, S. P. Pradhan, T. N. Singh, and P. G. Ranjith. 2013. Numerical modeling of Gondwana coal seams in India as coalbed methane reservoirs substituted for carbon dioxide sequestration. *Energy;* doi: 10.1016/j. energy. 2012.09.045

21. Vishal, V., P. G. Ranjith, S. P. Pradhan, and T. N. Singh. 2013. Permeability of sub-critical carbon dioxide in naturally fractured Indian bituminous coal at a range of down-hole stress conditions, *Engineering Geology* 167: 148–156

CHAPTER **17**

CO$_2$ Storage, Utilization Options, and Ocean Applications

M. A. Atmanand, S. Ramesh, S. V. S. Phani Kumar, G. Dharani, N. Thulasi Prasad, and Sucheta Sadhu
National Institute of Ocean Technology (Ministry of Earth Sciences),
Chennai-600100
E-mail: phani@niot.res.in

INTRODUCTION

The concentration of carbon dioxide (CO$_2$) in the atmosphere is promoted by the combustion of fossil fuels for power production and for industrial processes. Consequently, carbon capture and sequestration (CCS) from the point sources such as industries is necessary for meaningful greenhouse gas (GHG) reduction in the immediate future. The generally accepted and likely optimistic goal is to limit the global temperature increase to 2.1°C above pre-industrial levels by 2100 [1]. Intergovernmental Panel on Climate Change (IPCC) has estimated that it would require 50%–85% emission reduction from present levels by 2050 with emissions peaking no later than 2025 [2].

Capturing CO$_2$ from flue-gas streams of power plants is an essential parameter for the carbon management and sequestration of CO$_2$ from our environment [3]. Well-known physical and chemical adsorption of CO$_2$ for capture can be achieved by using solvents, cryogenic techniques, membranes, and solid sorbents. The large-scale operation of these technologies is energy intensive. Captured CO$_2$ is to be sequestered with suitable process for its long-term mitigation. Storage options include geological, biological, and ocean sequestration. This chapter discusses the perspectives of ocean sequestration of CO$_2$ based on the expertise and technology developed at the National Institute of Ocean Technology (NIOT) by means of various activities towards renewable and non-renewable ocean resources for coastal and island applications [4].

MAJOR SEQUESTRATION TECHNIQUES

Geological Sequestration

Geological sequestration involves pumping of CO_2 as liquid or critical fluid into the geological formations, aquifers, oil and gas reservoirs, where suitable porosity and permeability exists. The use of CO_2 pumping into the petroleum reservoir for enhancing the oil extraction from the reservoir is one of the options being implemented in some of the wells. The CO_2 reduces the viscosity of oil and acts as a driving force for pushing the oil to the wellbore. A total of 5 giga tonnes (Gt) CO_2 storage potential is available in India in the coal, oil, and gas fields [5]. Saline aquifer in the Ganges basin can also present geological storage capacity for long term. India has the capacity of 360 billion tonnes (Bt) CO_2 storage in deep saline aquifers [6]. Studies are in progress to utilize the vast Deccan Basalts as viable storage option, but the presence of deep-seated fractures controlled with tectonics necessitates the long-term behavioral studies of the storage.

Using CO_2 for coalbed methane (CBM) extraction is another option that is being explored with two prospective advantages, namely the compatibility of CO_2 sequestration in coalbeds and the storage of the gas mainly by adsorption. Storage in the deep-seated coal is being discussed as one of the options for the extraction of CBM.

Ocean Sequestration

Ocean water has long been identified as the natural sink or absorber for the atmospheric CO_2. However, the timescales for ocean processes are too long as compared to the rate of increase of the atmospheric CO_2. Impacts of the increase in atmospheric CO_2 content are identified as ocean acidification and the consequential impact on the natural carbon cycle. Two options for direct ocean sequestration are identified as (i) direct injection to different depths from 300 m to 1000 m and (ii) fertilization using iron filings and/or nitrogen. Some of these concepts were pursued for ocean sequestration, but stopped due to environmental concerns [7].

Disposal of CO_2 in deep ocean at depths of 2800 m below the sea level is another possibility. A locally stable pool of CO_2 under the heavy water column is created and expected to provide long-term storage. The technological feasibility of CO_2 transportation and disposal at these depths [8] has been studied in detail. The CO_2 can be effectively stored on the ocean floor with local pressure–temperature conditions of the ocean basins favouring the formation of a stable pool. However, environmental impact of any destabilizing effects due to natural disasters is not known.

CO_2 UTILIZATION IN OCEAN

Indirect processes of CO_2 storage in oceans utilize CO_2 for extraction of gas hydrates and in carbonation of industrial waste disposed in oceans.

Extraction of Gas Hydrates

The concept of CO_2 utilization by pumping CO_2 to replace methane (CH_4) in the gas hydrates [9–10] for their extraction has been proposed. ConocoPhilips project aims at such a demonstration at Ignik Sikumi field site in Arctic Circle. Advantages of this process besides extraction of hydrates with CO_2 [11] are contribution to CO_2 sequestration by replacing CH_4, and hydrate structure remains same and mechanical stability of the reservoir is not disturbed. CO_2 hydrates are thermodynamically more stable than CH_4 hydrates.

Carbonation Experiments

Utilization of CO_2 for carbonation of industrial wastes and construction of artificial reefs in the ocean have been proposed. CO_2 may be sequestered into stable compounds formed out of industrial wastes for safest and longest storage similar to the natural weathering process [12]. Numerous processes for mineral carbonation have been suggested and investigated with varying degrees of complexity [13]. The reactive nature of CO_2 with minerals and other waste materials needs to be understood for industrial wastes such as steel slag, fly ash, and pulp industry waste.

Indirect option of mineral CO_2 sequestration is a chemical sequestration route of which the formed products are thermodynamically stable and environmentally benign. Industrial wastes such as steel slag contain oxides of calcium, magnesium along with silica and can be considered as CO_2 storage options. These oxides react with CO_2 to form the carbonates that can be used for permanent and inherently safe storage. Energy efficiency of the carbonation process for the industry depends on the process and source material. The carbonated material could be used for making the

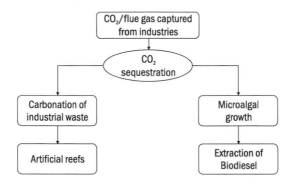

Figure 1 *Proposed methodology for the utilization of the sequestered CO_2 at NIOT*

structures for coastal protective measures, artificial reef growth, and so on. The formed mineral carbonates as end-products are known to be stable over geological time periods.

Research and Technology Development at NIOT

At NIOT, Chennai methodology has been proposed for CO_2 utilization in ocean (Figure 1). A high pressure reactor facility set-up has been developed at NIOT to work for carbonation experiments. The operating conditions are 100°C temperature and 20 bar pressure [12]. The laboratory set-up is shown in Figure 2.

In this industrial waste is ground to less than 75 micron size and exposed to CO_2 in a high pressure reactor at about 60°C for converting the oxides into the carbonates. Carbonated materials are then analysed in Scanning Electron Microscope and Fourier Transform Infrared Spectroscopy. Results for unreacted and reacted industrial waste are shown in Figures 3 and 4.

The carbonation process of the material is being analysed. Further experiments are underway with combination of acids/reactants for enhancing energy efficiency and test the effectiveness of conversion process.

FEASIBILITY STUDY TO USE CONVERTED CARBONATES AS COASTAL PROTECTION STRUCTURES

Carbonated wastes are proposed to be used in the preparation of the coastal structures. It has been documented in the literature that the

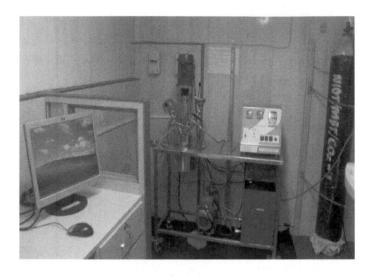

Figure 2 *Laboratory set-up for carbonation experiment at NIOT*

(a) Unreacted industrial waste   (b) Reacted industrial waste

Figure 3 *SEM analysis for unreacted and reacted industrial waste showing carbonation*

blocks made out of carbonated industrial wastes could be used for the artificial growth of corals in the sea. The depletion of beach sand is identified as one of the mechanisms leading to erosion and loss of reef crests and increased wave energy at the shore. Artificial reef for the coastal protective measures is one of the options and the institute has already shown the application of the artificial reefs to address the issue. Based on the converted carbonates, blocks are prepared to measure the leaching process in marine environment. Study in a shorter duration does not show much impact or changes in temporal leaching mechanism and pH.

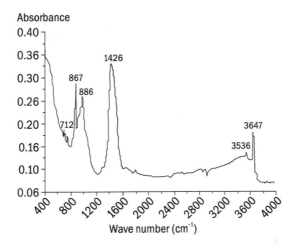

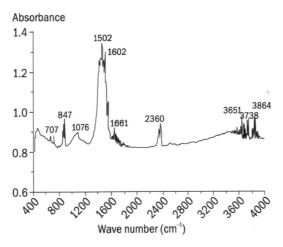

Figure 4 *FTIR analysis for unreacted (top) and reacted (bottom) industrial waste*

CO₂ SEQUESTRATION IN MARINE MICROALGAE

Biological sequestration involves fixing of CO_2 in biomass. The basic procedure involved in this process is fixation of CO_2 in the cultivation of microalgae. Microalgae could be cultivated with the captured CO_2 and used for biodiesel or nutraceuticals. NIOT is also involved in the bio-sequestration studies for identifying suitable marine algae for microalgal production of biodiesel and neutraceuticals. Pilot scale mass culture experiments are conducted and successfully optimized for high yielding strains with high lipid contents (Figure 5). Apart from natural fixation of microalgal, facilities are established for the introduction of microalgal

Isolation and maintenance of stock   Tubular Photobioreactor (600 L) Chennai.

Figure 5 *Algal production experiments at NIOT*

and flue gas for studying the microalgal growth patterns and tolerance levels in the presence of CO_2.

CONCLUSION

By comparing various options of CO_2 sequestration with reference to the Indian scenario for storage potential and the need for necessary industrial development, indirect ocean sequestration seems to be a potential option. Studies indicate that the carbonation from industrial wastes can be optimized for energy efficiency and utilized as artificial reef blocks to store the CO_2 for long term. Storage in marine environment as artificial reefs will be eco-friendly to the marine biotic life cycle. Option of CO_2 storage in microalgae will give added benefits for biodiesel production.

Summary

Sustained worldwide growth in population and economic activity has increased anthropogenic CO_2 emissions, which puts stress on the natural carbon cycle. The solutions put forth by various studies included storage and sequestration of CO_2 gas. In this chapter, brief outline of CO_2 storage options and perspectives in ocean disposal are discussed. Since, CO_2 is a global warming gas and there are alarming ecological and environmental concerns for direct disposal, especially into the ocean its indirect disposal by carbonation of industrial wastes and consequent usage for coastal protective measures such as artificial reefs is described. Microalgal growth for the extraction of biodiesel from CO_2 utilization in oceans is discussed in the current work.

ACKNOWLEDGEMENT

Authors would like to thank Ministry of Earth Sciences, Government of India for financial support and encouragement. Authors would also like to thank Visakhapatnam Steel Plant for inputs during the course of the studies and Dr P. S. R. Prasad, Scientist, NGRI, Hyderabad for carbonation analysis.

REFERENCES

1. Commission of the European Communities (CEC). 2007. Communication from the Commission to the Council, the European Parliament, the European Economic and Social Committee and the Committee of the Regions. *Limiting Global Climate Change to 2°C. The way ahead for 2020 and beyond.* Brussels, COM (2007) 2, final, EU, Brussels, Belgium

2. Metz, B., O. R. Davidson, P. R. Bosch , R. Dave, L. A. Meyer (eds). 2007. IPCC Climate Change 2007: Mitigation. Contribution of Working Group III to the Fourth Assessment Report of Intergovernmental Panel on Climate Change (IPCC), Cambridge University Press, Cambridge, UK

3. Gray, M. L., Y. Soong, K. J. Champagne, John Baltrus, Jr, R. W. Stevens, P. Toochinda, and S. S. C. Chuang. 2004. CO_2 capture by amine-enriched fly ash carbon sorbents. *Separation and Purification Technology* 35:31–36

4. Atmanand, M. A. 2013. Development in underwater technologies – Indian scenario. *Proceedings International Symposium on underwater technologies,* March, University of Tokyo, Tokyo, Japan

5. Holloway, S., A. Garg, M. Kapshe, A. Deshpande, A. S. Pracha, S. R. Khan, M. A. Mahmood, T. N. Singh, K. L. Kirk, and J. Gale. 2009. An assessment of the CO_2 storage potential of the Indian subcontinent. *Energy Procedia* 1:2607–2613

6. Garg, A. and P. R. Shukla. 2009. Coal and energy security for India: Role of carbon dioxide (CO_2) capture and storage (CCS). *Energy* 34:1032–1041

7. Blain Stéphane, B. Quéguiner, and L. Armand *et al.* 2007. Effect of natural iron fertilization on carbon sequestration in the Southern Ocean. *Nature* 446:1070–1074

8. Sarv, H. 1999. Large-scale CO_2 Transportation and Deep ocean Sequestration, phase I Final report. Prepared for US Department of Energy, National Energy Technological Laboratories, DE-AC26-98FT40412, March McDermott Technology Inc.

9. Ohgaki, K., K. Takano, H. Sangawa, T. Matsubara, and S. Nakano. 1996. Methane exploitation by carbon dioxide from gas hydrates-phase equilibria for CO_2-CH_4 mixed hydrate system. *J Chem Eng Jpn* 29:478–483

10. Hirohama, S., Y. Shimoyama, A. Wakabayashi, S. Tatsuta, and N. Nishida. 1996. Conversion of CH_4-hydrate to CO_2-hydrate in liquid CO_2. *J Chem Eng Jpn* 29(6):1014–1020

11. Le Quang Duyen, Jean-Michel Herri, Ouabbas Yamina, Truong Hoai Nam, and Le Quang Du. 2012. CO_2–CH_4 exchange in the context of CO_2 injection and gas production from methane hydrates bearing sediments. *Petroleum Exploration and Production* 10:38–45

12. Huijgen, W. J. J., G. J. Witkamp, and R. N. J. Comans. 2005. Mineral CO_2 sequestration by steel slag carbonation. *Environmental Science and Technology* 39:9676–9682

13. Olajire, A. A. 2013. A review of mineral carbonation technology in sequestration of CO_2. *Journal of Petroleum Science and Engineering* pp. 364–392

Index

A

Accelerated Power Development & Reform Programme (APRDP), 14

Activated carbon, 76–78, 84, 95–97
  properties of, 77t

Adsorption isotherms, 251

Al Gore, 189
  An Inconvenient Truth, 189

Algal sequestration of CO_2, 169

Ankleshwar oilfield, 124, 129

Annex I countries, 3, 7

Australia, 6, 7, 9, 32, 35–41, 49, 50, 53, 58, 86, 115, 134, 199, 238, 246
  CCS projects of, 39–41
  major initiatives for, 41–42

B

Basalt formations, 34, 123, 125, 126, 129
  main CO_2 trapping mechanisms in, 125

BHEL, 11

Bio-carbon capture and storage (Bio-CCS), 127, 129, 130

Bioflocculants, 215

Biomimetic carbon sequestration, 167–182

Bio-sequestration of CO_2, 34, 52, 127, 129, 167, 168, 189–201
  microalgae in, 191–192
  potential and challenges, 189–201
  through photosynthesis, 168

Biosurfactants, 207–221

Bituminous coal, 78
  gas composition for, 78t

Bureau of Energy Efficiency (BEE), 21, 29

C

Calvin cycle, 209

Captured CO_2, 4, 31, 33, 34, 35, 40, 45, 50, 57, 65, 226, 230, 257, 262
  transportation and injection of, 33, 40
  underground geological storage of, 33–34
  utilization of, 35, 45, 50

Capturing carbon dioxide 9, 10, 31–53, 57–66, 69–88, 95–96, 99, 101–104, 123, 173–177, 189, 212, 218, 223–230, 243
  by adsorption, 75f, 75–77
    from flue gas, 77–78
    technologies for, 96
  aqueous alkanolamine solutions for, 71–73
  by carbonate process, 73
  by chemical absorption, 70–71, 71f
  by chilled NH_3 process, 73–74
  by chemical looping combustion (CLC), 80f, 80–81
  energy requirements for, 74t
  fixation technologies for, 70f
  from flue-gas, 257
  gas processing prior to, 85–86
  global R&D and industrial facilities for, 86–87
  importance of, 133
  metal organic frameworks (MOFs) in, 99–100

by oxy-fuel technology, 78, 80

by physical and chemical means, 69–88

from post-combustion flue gas, 96–99

from power plants, 75–76, 95–104

post-combustion technologies for, 69–81, 95

pre-combustion technologies for, 69, 81–86, 95

reactor unit of, 173*f*

role of forests in, 149–160

under water gas shift reactor, 101–102

Carbon capture and storage (CCS), 3–11, 16, 32, 39, 51–54, 57–67, 93, 120, 124, 127, 147, 150, 182, 189, 211, 238, 254

adoption of, 17–18

awareness and capacity building in, 51–52

carbon recycling through, 232–235

in coal seams, 254

cost of, 8–9

earth processes, 52

explained, 57

implementation in India, 61–66
see also India

barriers to, 63–64

capacity development needs of, 65

recommendations for, 66

limitations of, 6–7

scoping study in India, 58–60

technology, 6, 16–18, 39, 42, 45, 171, 211

options, 9, 10*t*

three major aspects of, 57–58

Carbon capture and storage (CCS) projects, 39–46, 51, 53

capacity building in, 51–52

in Australia, 39–41
Australia

in India, 46–48
see also India

in People's Republic of China, 44–46

see also People's Republic of China

and research publications, 48–50

comparison of, 50*f*

future steps, 50–51

growth in, 49–50, 49*f*

number of, 51*f*

in South Africa, 43–44
see also South Africa

in United States of America, 41–43
see also United States of America

Carbon capture and storage and utilization (CCSU) technologies, 32, 39–40, 48, 52

current technologies, 39–40

sun sets of, 32–35

Carbon dioxide (CO$_2$),

emissions, 4–8, 16, 18, 19, 25–28, 32, 35–47, 53, 58, 60, 61, 69, 107, 108, 121, 123–131, 157, 168, 189, 210–212, 223, 243, 245, 263

in 2012, 7*t*

and CCS supporting policies, 40, 40*f*

global trends of, 6, 36*t*, 37*f*

reduction commitment of, 38*t*

geological storage of, 108–110, 123–130

see also Geological storage of CO$_2$

geothermal power from, 128, 129*f*

and methane absorption, 140

mineralization of, 178–179

reduction, 19–22

measures for, 21–22

strategies for, 19

storage,

    in depleted oil reservoirs, 123–124

    options for, 123–125

utilization in ocean, 259, 260, 263

Carbon dioxide (CO_2) sequestration, 31, 34, 35, 45, 47, 51–56, 61, 82, 122, 126, 129, 133–147, 176, 178, 182, 225, 226, 229f, 229, 230, 252, 258, 259, 262, 263

    of Indian coal fields, 133–147

    in marine microalgae, 262–263

Carbon fixation, 168, 201, 209, 210, 213

    microbes in, 168–169

    in terrestrial ecosystem, 34–35

Carbon sequestration, 3, 35, 39, 42, 47, 48, 61, 116, 126, 149–160, 167–182, 189, 192, 195, 196, 198, 245, 249–255

    biomimetic, 167–182

    in coal seams, 249–255

        implications of, 249–255

    in Manipur forests, 155t

    heterotrophic bacteria in, 170

        see also Heterotrophic bacteria

    rate of, 155

    in soil, 155

    potential in North-eastern India, 149–160

        see also North-eastern India

Carbon Sequestration Leadership Forum (CSLF), 3–5, 9, 39, 42, 43, 46, 47, 52

    financing mechanism, 5

    ministerial meetings, 4–5

    number of projects, 4

    objective, 4

technology road maps of, 9

Carbonation, 34, 35, 104, 179, 185, 186, 259–261, 263–265

    of industrial wastes, 259, 263

Carbonic anhydrase, 34, 170–172, 177, 179, 181, 182

    bio-mineralization with, 178

    carbon capture process of, 171–178, 171–178

Central Electricity Authority (CEA), 15, 16, 19

Central Electricity Regulatory Commission (CERC), 15, 21, 22, 25, 29

Centre for Power Efficiency and Environmental Protection (CENPEEP), 46, 231

Chemical looping combustion (CLC), 10, 80–81

    see also Capturing carbon dioxide

Clean coal technologies mission, 59

Clean development mechanism (CDM), 43, 158, 225

CO_2-EOR, 123, 124, 129, 130, 242, 246

Coal India, 11

Coal matrix deformation, 251–252

    CO_2 sorption-induced, 251

Coal permeability, 109, 250–252, 255, 256

    cleats and matrix elements in, 252f

    swelling-related loss in, 252

Coal production, 35–36

    global increase in, 36

    share in energy, 36t

Coal resources, 35, 46, 136

    estimates of, 35, 46, 136t

    in India, occurrence of, 133

Coal seam sequestration, 249, 250, 254, 255

and reservoir behaviour, 250
in India, 254, 255
Coal-based economies, 31–53
  coal production data of, 36t
  coal-fired plants in 2010, 37t
  emission reduction commitment of, 38t
  emissions trends in, 36t, 37f
Coalbed methane (CBM), 63, 108, 109, 123, 124, 133, 250, 258
  extraction, 258
  recovery, 109, 133
  resource in India, 253
Coalfields, 34, 133–147, 253
  concealed, 139, 141, 143
    analysis and rank of, 142t
    CO$_2$ storage capacity of, 145t
  grey areas of, 137
  in India, 133–147, 253
  well-delineated, 139
Converted carbonates, 260, 261
  as coastal protection structures, 260–261
Copenhagen accord, 42, 43
CSIR-IIP, 96, 97, 100
  PVSA unit at, 97f
Cyanobacteria, 128, 169, 172, 178, 207–218
  advantages of, 212–213
  as biosurfactants and flocculants, 217–221
  in carbon capture, 207–218
    see also Capturing carbon
  described, 207–208
  extracellular polymeric substances in, 213–215
  metabolic pathways of, 210f

D
Dakota Gasification Company (DGC), 241

Darcy's law, 251, 252
Depleted oil reservoirs, 123
  see also Carbon dioxide (CO$_2$)

E
East Bokaro Coalfield, 138, 141t, 143, 144t
ECBM recovery, 45, 50, 108, 109, 125, 133, 143, 251, 254
Electricity Act 2003, 14
Electricity generation, 8, 21, 58, 59, 65, 169,
  coal-based, 6, 9, 12, 41, 44, 46, 47, 59, 224, 229
  India's installed capacity of, 59
Enhanced coalbed methane (ECBM) recovery, 35, 45, 50, 51, 108, 106, 124, 125, 129, 143, 250, 251, 254
  factors affecting, 125
Enhanced oil recovery (EOR), 45, 60, 86, 107, 109, 123, 237–248
  case studies on, 239–248
  monitoring/flow surveillance in, 244–248
  technology, 237–248
Environmental Protection Act, 1986, 63
Enzymatic carbon capture, 171–172

F
Fischer-tropch synthesis, 33
Flocculants, 200, 207, 215, 218
  see also Cyanobacteria
  defined, 215
Flue Gas Desulphurization (FGD), 75, 194
Flue gas regeneration, 224
  chemistry of, 224–225
FTIR analysis for industrial waste, 262f

G
Geo-Green concept, 127

Geological sequestration, 42, 54, 107–121, 249, 258
  defined, 249
Geological storage of CO_2, 108–110, 123–130
  see also Carbon dioxide
  in basalt formations, 125–126
  in coal seams, 109, 124–125
  in deep saline aquifers, 109–110
    advantages of, 110
    distribution of, 115–116, 116*f*, 116–119
    issues connected with, 112–113
    mechanism of, 110–111
    screening criteria for, 113–115
  in deep underground formations, 108*f*
  in depleted oil and gas reservoirs, 108
  in geochemical trapping, 112
  Indian initiatives for, 129
  in oceans, 109–110
  options for, 108–109
  in physical trapping, 111–112
  in saline aquifers, 126–127
Geological Survey of India, 46, 126, 135, 137, 142, 147
Geostratiphic acceptance, 33
Gigawatt scale power plant, 61, 62
Global Carbon Capture and Storage Institute (GCCSI), 39, 40, 41, 44, 47, 52, 58
Global Environmental Facility (GEF), 158
Global warming, 51, 59, 69, 95, 120, 167, 168, 181, 182, 241, 249, 263
Gondwana formation, 133, 134, 137, 139
  in different states of India, 136*t*
Great Plains Synfuels Plant, 241

Green Energy Technology Center (GETC), 225
Greenhouse gas (GHG), 3, 17, 37, 41, 45–47, 52, 57–59, 65, 69, 86, 120, 123, 127, 146, 149, 167, 168, 189, 193, 195, 200, 201, 209, 210, 223, 224, 243, 249, 250, 254, 257
  control of, 149, 249
  effects of, 167
  increase in, 209
  sources of, 123
Greenpeace, 8
Grey area coalbeds, 141, 143, 144
  CO_2 adsorption capacity in, 144*t*
Grid connectivity, 20

H
Heterotrophic bacteria, 167, 170
  see also carbon sequestration
  in carbon sequestration, 167, 170
  non-photosynthetic CO_2 fixation by, 170–171
  see also Non-photosynthetic CO_2 fixation
Hydrogen membrane reforming (HMR), 33, 81, 82
  schematic diagram of, 82*f*

I
ICOSAR, 47
In-combustion processes, 32–33
India Greenhouse Gas Inventory 2007, 47
India Smart Grid Forum, 28
India Smart Grid Task Force, 28
India,
  candidates for CO_2 storage in, 141*t*
  CO_2 emissions data for, 58
  CO_2 emitting sectors of, 59–60
  coal and lignite deposits in, 134–135, 135*f*

coal inventory of, 135–136

current CCS activity in, 61–62

current climate change policies of, 58–59

economic analysis of, 61–62

extraneous coal deposits of, 136–137

oil and gas production in, 59–60

policy and legislation review in, 62–63

  environmental impact assessment, 63–65

  for groundwater, 63

  for oil and gas, 62

potential coalbeds for CO_2 Storage in, 140–143

  properties of, 140–141

  storage capacity of, 142–143

power sector in, 13–19, 23, 24, 29, 59

  CO_2 baseline data for, 16–17, 16t

  efficiency measures for, 19

  electricity installed capacity in 2011

  evolution of, 13–29

  future capacity addition for, 23, 24t

  generation capacity of, 9, 11, 12, 13t, 59

  growth in capacity, 273

  steps for CO_2 abatement in, 18–19

  transmission and distribution of, 15–16

Indian cement industry, 60

Indian iron and steel industry, 61

Indian Petroleum Act, 1934, 62

Indigenous Forest Management, 158

Integrated Energy Policy (2006), 9

Integrated Gasification Combined Cycle (IGCC), 10t, 16, 33, 45, 70, 80, 81, 84, 85, 87, 88, 223

  benefits of, 84–85

  process flow diagram of, 84f

Intergovernmental Panel on Climate Change (IPCC), 4, 8, 37, 150, 153, 189, 257

International Energy Agency (IEA), 36, 42, 43, 51, 52, 55, 57, 120, 127, 246

Ishikari basin, 250, 251

J

Jharia coalfield, 136, 136t, 137, 139, 141t, 143, 144t,

K

Kyoto Protocol, 3–5, 38t, 39, 149, 189, 200, 210, 243

  Article 3.3, 149

  ratified, 3, 38, 39, 43, 189

  target of, 3, 47, 243

L

Lake Nyos leakage, 8

M

Membrane Reforming Sorption Enhanced Water-Gas-Shift (MRSEWGS), 70

Membrane water-gas-shift reaction (MWGSR), 81, 88

Metal organic framework, 96, 99, 105

Methanogens, 170, 171, 181

Microalgae mass culture, 191–192

Microalgal species, 191f

  biotechnological uses of, 193t

  CO_2 mitigation by, 192–195

  cultivation with flue gases, 195–200

    challenges in, 196–200

  features of, 192

  photosynthesis in, 212f

Ministry of Earth Sciences, 64

Ministry of Science and Technology, 45, 47, 61

Multi-Purpose Fuels, 223–231

N

National Action Plan on Climate Change (NAPCC), 18, 19, 47, 58, 150

National Aluminium Company (NALCO), 61

National Clean Energy Fund, 11

National Commission on Energy Policy, 275

National Institute of Ocean Technology (NIOT), 257, 260–263
  algal production experiments at, 263*f*
  methodology of sequestered CO_2 at, 260*f*
  research and technology development at, 260–263

National Mission for a Green India, 18

National Mission for Enhanced Energy Efficiency (NMEEE), 18, 19, 22, 46, 59

National Mission for Sustainable Agriculture, 18, 59

National Mission for Sustaining the Himalayan Ecosystem, 18, 59

National Mission on Enhanced Energy Efficiency (NMEEE), 19, 22, 46

National Mission on Strategic Knowledge for Climate Change, 18, 59

National Mission on Sustainable Habitat, 18, 59

National Programme on Carbon Sequestration (NPCS) Research, 47, 61, 150

National Solar Mission, 18, 47, 59

National Thermal Power Corporation (NTPC), 11, 61, 96,

National Water Mission, 18, 59

Non-Annex I countries, 3, 47, 56

Non-photosynthetic CO_2 fixation, 170, 182
  see also Heterotrophic bacteria
  by Heterotrophic bacteria, 170

North-eastern India, 149–160
  see also Carbon sequestration
  biomass and carbon density in, 154t
  carbon sequestration potential of, 149–160
  carbon stock in, 152–153
  forest cover of, 151, 151*t*–152*t*
  forest management practices in, 158
  organic carbon content in the soils of, 156*t*
  physiogeographic zone of, 152
  rate of carbon sequestration in, 155t, 155–156
  soil CO_2 emissions in, 157–158, 157*t*
  state-wise growing stock in forests in, 153*t*, 152–154

Northern Regional Load Dispatch Center (NRLDC), 22

Nuclear power, 6

O

Ocean fertilization, 190

Ocean sequestration, 167, 257, 258, 263, 264
  of CO_2, 257

Oil and Natural Gas Corporation (ONGC), 11, 61, 120, 124, 247

Oil Industry (Development) Act, 1974, 62

Oilfields (Regulation and Development) Act, 1948 62

Oxy–fuel combustion, 45, 78, 79, 90, 95, 104, 224
  advantages of, 78–79
  flow sheet of, 79*f*

P

Palladium-ceramic membrane, 82
    process flow diagram of, 83f
PAT Energy Efficiency Certificate, 22
PAT scheme, 22–23
    energy conservation under, 23t
People's Republic of China (PRC), 32,
    35–39, 42, 44–46, 49, 50, 53
    CCS projects in, 32, 44–46, 53
    coal consumption in, 35
    Coal-fired plants in, 37t
    coal production in, 36t, 36
    coal reserves of, 44
    emission reduction commitment
        of, 38t
Permian oil basin, 238–239
Petroleum and Natural Gas Rules,
    1959
Petroleum Mineral Pipelines (Acquisition
    of Right of User in Land) Act, 1962,
    62
PETRONAS plant, 72
Photosynthesis, 34, 87, 127, 149, 167–
    169, 181, 190, 192, 195, 207, 209,
    211, 212
    in algae, 212f
    role of microbes in, 169
Post-combustion capture, 10, 32, 69,
    88, 104
    see also Capturing carbon dioxide
    (CO_2)
Pre-combustion capture, 33, 69, 70, 81,
    87
    see also Capturing carbon dioxide
    (CO_2)
Pressure (or vacuum) swing adsorption,
    32, 75, 87, 89, 90, 96
Pressure and temperature swing
    adsorption (PTSA), 78, 102
Pressure-volume swing adsorption
    (PVSA) cycle, 96, 97, 104
    step sequences of, 98t

R

Rajiv Gandhi Grameen Vidyutikaran
    Yojna, 14
Rajiv Gandhi Technological University
    (RGPV) study, 223, 225–230, 280
    methodology and specifications of,
        226–227
    objectives of, 226–227
    results of, 227–228
    schematic diagram of, 227f
Raniganj coalfield, 137–139, 141t, 143,
    144t, 254, 280
R-APDRP, 14, 27
Renewable Energy Certificates (REC),
    21, 22, 29
Renewable energy, 6, 9, 14, 20, 21,
    23–29, 46, 47, 167, 211, 223, 225
    Contribution in capacity addition,
        24f
    contributions of, 24f
    development, 20–21, 23
    integration, major Challenges in,
        24–25
    in RE-rich states, 25t
    management centers, 27–28
    purchase obligation, 21, 22
    sources, 6, 9, 12–14, 20–29, 46, 47,
        58, 167, 180, 200, 211, 215, 223,
        257
Reuter, Thomson, 48

S

SEM analysis for industrial waste, 261f
Single column microadsorber unit, 100,
    100f
Skarstrom cycle, 96, 98
Sleipner project, 34, 86, 107, 115, 126,
    243
Smart grid, 27, 28
    integration of, 27
Sohagpur coalfield, 138, 141t, 143, 144t

Soil organic carbon (SOC), 155–159, 168

Sorber Enhanced Water-Gas-Shift (SEWGS), 81, 83

Sorption enhanced water-gas-shift, 70, 83

South Africa, 32, 35–39, 43, 44, 50, 53, 134
   CCS projects in, 32, 35, 39, 43–44, 50, 53
   coal production in, 36*t*
   coal consumption in, 35
   coal–fired plants in, 37*t*
   coal reserves of, 43
   emission reduction commitment of, 38*t*, 43, 53,

South African National Energy Research Institute (SANERI), 44

South Karanpura coalfield, 138, 141*t*, 143, 144*t*

State Forest Management, 158

Supervisory Control AND Data Acquisition (SCADA), 14, 27

T

Talcher coalfield, 138, 141t, 142, 143t, 144*t*

Temperature Swing Adsorption (TSA), 75, 78, 102

Terrestrial Sequestration, 34, 52, 190

U

Undersea gas hydrates, 34, 259
   extraction of, 259, 260

United Nations Framework Convention on Climate Change (UNFCCC), 3, 38*t*, 43, 149

National Communications by India to, 3

United States of America (USA), 3, 6, 7, 17, 32, 35–42, 44, 47,49, 53, 81, 86, 110, 112, 114, 115, 126, 192, 237–239, 241, 243, 246, 250
   case study from, 239
   CCS projects in, 17, 32, 39, 42, 47, 49, 53
   coal production in, 36*t*,
   coal-fired plants in, 37*t*
   emission reduction commitment of, 38, 42
   EOR technologies in, 237, 239
   oil production in, 238

Unmineable coalbeds, 139, 141, 149
   analysis and rank of, 141*t*
   and CO_2 storage capacity,143*t*
   and mineable coalbeds, 142

W

Web of Science, 48–50, 53

Weyburn oilfield, 124, 240*f*, 240–243
   oil recovery in, 240–243
   well grid structure in, 241*f*

Weyburn project, 42, 86, 107, 124, 240–243

Weyburn-Midale project, 124

Wide area management systems, 27

Wind variability in Rajasthan, 25–26, 26*f*

Z

Zeolites, 76, 77, 95, 96, 104

Zero Emission Future Gen Project, 5

About the Editors

Malti Goel, M. Sc. (Physics) from BITS, Pilani and D.I.I.T. PhD from IIT, Delhi, was the first Indian scientist to represent India as Vice-chair to the Technical Group of an International Forum on Carbon Sequestration Leadership. As Adviser and Scientist 'G', Ministry of Science and Technology, Government of India, she headed the Inter-sectoral Science and Technology Advisory Committee Division (1998–2008). She has been Emeritus Scientist, Indian National Science Academy and in the Centre for Studies in Science Policy, Jawaharlal Nehru University, New Delhi. Recipient of several awards and honours, she was conferred *Bharat Jyoti Award* for her outstanding contribution to science and society in 2012. She became Fellow of National Environment Science Academy in 2008. She has been invited by Subsidiary Body of Science and Technology Advice of UNFCCC and also by other International Academies to deliver lecture on S&T issues in climate change. Dr Malti Goel has published over 101 research papers in the journals of international repute and presented more than 150 papers in conferences. She has made important contribution through books on emerging environmental topics such as *Energy Sources* and *Global Warming (2005); Weather and Climate (2007); Carbon Capture and Storage R&D Technologies for Sustainable Future (2008); CO$_2$ Sequestration Technology for Clean Energy* (2010) as well as *Urja Avem Carbon Dioxide; Ikkisvi Shatabdi ki Chunotiyan* (2012).

M. Sudhakar, M. Sc. Tech. and PhD from Indian School of Mines, Dhanbad, as a British Council Fellow has a Master's degree in Law of the Sea and Marine Policy (1990) from the London School of Economics and Political Science, UK. He has served in premier research institutions of the country, such as National Institute of Oceanography and National Centre for Antarctic and Ocean Research (1997–2009) in Goa and contributed extensively for research and development in the field of Oceanography/Offshore Surveys/Polar Sciences/ Marine Technology. Since 2009 he has been serving as an Adviser and Scientist 'G' in the Ministry of Earth Sciences, Government of India, New Delhi and is heading the Awareness and Outreach Programmes and Vessel Management Divisions.

Recently he took over as the Director, Centre for Marine Living Resources and Ecology, Kochi under the MOES. He has published over 60 research papers in referred International/National Journals and has presented an equal number in conferences and seminars. He has been a referee of scientific journals and as a Guest Editor contributed a special section on "India's contribution to Southern Ocean" in *Current Science*. Dr Sudhakar has edited books such as *Scientific and Geopolitical Interests in Arctic and Antarctic, Management of Water, Energy and Bio-resources in the Era of Climate Change: Emerging Issues and Challenges*, and *Management of Natural Resources in a Changing Environment*.

R. V. Shahi did his graduation in Mechanical Engineering, post-graduation in Industrial Engineering, and post-graduate diploma in Business Management. He is Fellow of World Academy of Productivity Sciences, Institution of Engineers (India), International Institute of Electrical Engineers, and Indian National Academy of Engineering. From April 2002 to January 2007, he was Secretary to the Government of India, Ministry of Power. He was also President of the Governing Council of Central Power Research Institute, Chairman of the Executive Committee of Bureau of Energy Efficiency, and Chairman of the Governing Council of National Power Training Institute. R. V. Shahi is recipient of many awards and honours, including Powerline Expert Choice Award (2006) for Biggest Individual Contribution to the Power Sector on the basis of a survey of power professionals. He has contributed and presented many papers at various national and international conferences. Apart from editing *100 Years of Thermal Power in India*, he has authored many books such as *Indian Power Sector – Challenge and Response, Towards Powering India: Policy Initiatives and Implementation Strategy*, and *Energy Security and Climate Change*. His last book *Light at the End of the Tunnel? Way Forward for Power Sector* was released in 2013.

Printed and bound by CPI Group (UK) Ltd, Croydon, CR0 4YY

17/10/2024

01775681-0008